S(
Roberto Togneri
Anthony Zaknich

Auditory Features for Speech Recognition and Enhancement

Serajul Haque
Roberto Togneri
Anthony Zaknich

Auditory Features for Speech Recognition and Enhancement

Psychoacoustics of human hearing applied to automatic speech recognition

VDM Verlag Dr. Müller

Impressum/Imprint (nur für Deutschland/ only for Germany)

Bibliografische Information der Deutschen Nationalbibliothek: Die Deutsche Nationalbibliothek verzeichnet diese Publikation in der Deutschen Nationalbibliografie; detaillierte bibliografische Daten sind im Internet über http://dnb.d-nb.de abrufbar.

Alle in diesem Buch genannten Marken und Produktnamen unterliegen warenzeichen-, marken- oder patentrechtlichem Schutz bzw. sind Warenzeichen oder eingetragene Warenzeichen der jeweiligen Inhaber. Die Wiedergabe von Marken, Produktnamen, Gebrauchsnamen, Handelsnamen, Warenbezeichnungen u.s.w. in diesem Werk berechtigt auch ohne besondere Kennzeichnung nicht zu der Annahme, dass solche Namen im Sinne der Warenzeichen- und Markenschutzgesetzgebung als frei zu betrachten wären und daher von jedermann benutzt werden dürften.

Coverbild: www.purestockx.com

Verlag: VDM Verlag Dr. Müller Aktiengesellschaft & Co. KG
Dudweiler Landstr. 99, 66123 Saarbrücken, Deutschland
Telefon +49 681 9100-698, Telefax +49 681 9100-988, Email: info@vdm-verlag.de
Zugl.: Perth, University of Western Australia, Diss., 2008

Herstellung in Deutschland:
Schaltungsdienst Lange o.H.G., Berlin
Books on Demand GmbH, Norderstedt
Reha GmbH, Saarbrücken
Amazon Distribution GmbH, Leipzig
ISBN: 978-3-639-18396-2

Imprint (only for USA, GB)

Bibliographic information published by the Deutsche Nationalbibliothek: The Deutsche Nationalbibliothek lists this publication in the Deutsche Nationalbibliografie; detailed bibliographic data are available in the Internet at http://dnb.d-nb.de .

Any brand names and product names mentioned in this book are subject to trademark, brand or patent protection and are trademarks or registered trademarks of their respective holders. The use of brand names, product names, common names, trade names, product descriptions etc. even without a particular marking in this works is in no way to be construed to mean that such names may be regarded as unrestricted in respect of trademark and brand protection legislation and could thus be used by anyone.

Cover image: www.purestockx.com

Publisher:
VDM Verlag Dr. Müller Aktiengesellschaft & Co. KG
Dudweiler Landstr. 99, 66123 Saarbrücken, Germany
Phone +49 681 9100-698, Fax +49 681 9100-988, Email: info@vdm-publishing.com

Printed in the U.S.A.
Printed in the U.K. by (see last page)
ISBN: 978-3-639-18396-2

This book is dedicated to our family members, especially

Selina and Sajidul

for their support and inspirations.

Contents

Preface   ix

List of Tables   xiii

1 Introduction   1
   1.1 Methodology   5

2 Review of auditory perceptual features and speech recognition   9
   2.1 The peripheral auditory system   9
   2.2 Basilar membrane transduction and the critical bands .   12
        2.2.1 Rate-place theory   15
        2.2.2 Temporal-place theory   15
   2.3 Psychoacoustics   16
        2.3.1 Equal loudness curves of human perception . . .   17
        2.3.2 Half-wave rectification   19
        2.3.3 Compressive nonlinearity   19
        2.3.4 Synaptic adaptation   20
        2.3.5 Auditory masking   21
        2.3.6 Two-tone suppression   22
   2.4 Articulatory phonetics   23
   2.5 Computational models for speech processing   25
   2.6 Acoustic pre-processing and speech parametrization . .   27
        2.6.1 Static parameters   28

2.6.2 Perceptual linear prediction (PLP) 33

2.6.3 Temporal filtering and dynamic parameters . . 34

2.7 Automatic speech recognition 36

2.7.1 Deterministic speech recognition-pattern classifiers 36

2.7.2 Acoustic modelling using HMM 38

2.7.3 HMM for word recognition 39

2.7.4 Recognition using Viterbi decoding 44

2.8 Auditory models for speech recognition 46

3 Synaptic adaptation in a temporal auditory model   55

3.1 Temporal processing of speech with zero-crossing algorithm 60

3.1.1 The temporal adaptation model 65

3.1.2 Feature extraction with synaptic adaptation . . 70

3.2 CV (consonant-vowel) discrimination by temporal auditory adaptation 73

3.2.1   A measure of CV discrimination by synaptic
adaptation 74

3.2.2 Vowel clustering with synaptic adaptation . . . 75

3.2.3 Linear discriminant analysis of synaptic adaptation by class separability measure 77

3.3 HMM word recognition with synaptic adaptation . . . 79

3.4 Summary and conclusions 83

4 Speech recognition with two-tone suppression   85

4.1 Two-tone suppression in the auditory system 85

4.2 The companding architecture 86

4.3 Implementation of temporal companding using zero-
crossing algorithm 88

4.4   A comparative analysis of CV discrimination with synaptic adaptation and two-tone suppression 94

4.5   HMM word recognition with two-tone suppression . . . 95

4.6   Summary and conclusions 97

5   Frequency dependent asymmetric compression in speech recognition   99

5.1   Introduction 99

5.2   The Fletcher-Munson perception curves and equal loudness coefficients 102

5.3   Asymmetric compression in ASR parametrization . . . 104

5.3.1   Static logarithmic compression in the MFCC and the ZCPA 104

5.3.2   The compression ratio and the compression angle 106

5.3.3   The asymmetric compression coefficient 107

5.3.4   Application of asymmetric compression in the MFCC 108

5.3.5   Derivation of the compression angle 109

5.4   HMM word recognition with asymmetric compression in the MFCC 112

5.5   Summary and conclusions 114

6   Feature extraction for ASR based on higher auditory processing   115

6.1   The proposed CN processing strategy applied to ASR front-end 117

6.2   The central auditory system and the cochlear nucleus . 119

6.3   The anteroventral cochlear nucleus (AVCN) 122

6.4   The dorsal cochlear nucleus (DCN) 124

6.5   The posteroventral cochlear nucleus (PVCN) 124

6.5.1   The half-wave rectification in the synapse processing 125

6.5.2   The temporal synaptic adaptation - onset neurons 125

6.5.3   The mean (average discharge) rate processing . 126

6.6   The CN feature extraction model 127

6.7   The HMM word recognition with CN features 129

6.8   Noise considerations in the mean rate (MR) processing 130

6.9   Performance comparisons between CN features and MFCC features 134

6.10 Summary and conclusions 135

7   Psychoacoustic spectral subtraction methods for speech recognition   137

7.1   Computation of the auditory masking threshold (AMT) 141

7.1.1   The power spectrum 141

7.1.2   The critical band analysis 142

7.1.3   Adaptation (temporal forward masking) 143

7.1.4   Equal loudness pre-emphasis and cube-root compression 144

7.1.5   The spreading function 144

7.1.6   The auditory masking threshold (AMT) 145

7.2   Estimation of the perceptual noise 146

7.3   The T-F noise subtraction algorithm based on the AMT 148

7.4   HMM word recognition with psychoacoustic spectral subtraction 153

7.5   Summary and conclusions 155

8   Major findings and future research   157

8.1   Major findings 157

8.2   Future research 159

8.3   Generalized conclusions 160

Appendix A: A Statistical significance test for speech recognition algorithms.   162

Appendix B: Analysis of two-tone suppression by companding. 165

Appendix C: **167**
     C.1 Normalized equal-loudness weighting coefficients (L_m) for 40 channel mel-frequencies. 167
     C.2 Critical Bands and the Bark frequency scale. 167

Bibliography   **169**

Index   **183**

Preface

This book is the outcome of the doctoral research carried out at the Signal and Information Processing (SIP) Lab of the University of Western Australia during 2004-2008. Automatic speech recognition (ASR) is one of the most important research areas in the field of speech technology and research. It is also known as the recognition of speech by a machine or by some artificial intelligence. However, in spite of focused research in this field for the past several decades, robust speech recognition with high reliability has not been achieved as it degrades in presence of speaker variabilities, channel mismatch conditions, and in noisy environments. The superb ability of the human auditory system has motivated researchers to include features of human perception in the speech recognition process. The principle objective of this book is investigation of the roles of perceptual features of human hearing in automatic speech recognition in clean and noisy environments. A second objective is to identify the perceptual features which are relevant to the enhancement of automatic speech recognition performance.

The book is arranged in 8 Chapters and is organized as follows:

The first two chapters are largely introductory. Chapter 1 is an introduction to automatic speech recognition (ASR) and some problems of ASR in the current state of technology, and methodologies adopted in this work in addressing those problems. Chapter 2 includes a review of the auditory system and the psychoacoustic features, the state-of-the-art recognition techniques applied to ASR, including literature review and background research up to date on perceptual features for automatic speech recognition. The next five chapters, 3-7, detail the research and experimentations undertaken and implemented under this project and also summarizes the results and findings from these experimentations.

In Chapter 3, the effects of auditory adaptation on ASR are investigated. The perceptual properties of rapid and short-term synaptic adaptation are implemented in an auditory model utilizing temporal processing of speech using a zero-crossing algorithm, and the recognition performances are evaluated by a continuous density HMM recognizer in a speech recognition front-end. The role of synaptic adaptation in CV discrimination, and vowel clustering by grouped scatter plots are presented. The performance of synaptic adaptation in vowel discrimination using a class separability measure based on linear discriminant analysis and between-class and within-class scatter matrices is also developed.

In Chapter 4, the perceptual property of two-tone suppression is implemented in a ZCPA auditory model utilizing the temporal companding algorithm. The ASR performance of the two-tone suppression is compared with synaptic adaptation and the relative performances are evaluated in Gaussian and non-Gaussian noise environments.

Chapter 5 investigates the effects of static compression on ASR parametrization as observed in the psychoacoustic input/output (I/O) perception curves. Two frequency dependent compression strategies are introduced in the MFCC parametrization. The first is the equal loudness coefficients corresponding to the loudness versus frequency curves of human perception. In addition, a frequency dependent asymmetric compression technique, that is, application of higher compression in the higher frequency regions than the lower frequency regions is presented.

In Chapter 6, an approach to integrate an ASR front-end with the higher auditory processing strategies is developed. A method of feature extraction for ASR based on the processing in the cochlear nucleus (CN) is presented. The processing of synchrony detection, average discharge rate (synaptic processing) and the two-tone suppression corresponding to the AVCN, the PVCN and the DCN regions of the CN, respectively, are segregated and processed separately at the feature extraction level according to the differential processing scheme as observed in the CN. Central to the proposed system is the separation of the synchrony detection from the synaptic processing in the recognition algorithm.

A time-frequency spectral subtraction method based on several psychoacoustic properties of human perception is developed in Chapter 7. For speech corrupted with wideband noise, a tonal excitation pattern and an auditory masking threshold (AMT) is determined based on critical band filtering, synaptic adaptation which also introduces temporal forward masking, equal loudness pre-emphasis, power law compression of hearing, and simultaneous masking effect introduced by a spreading function. Noise is estimated considering that it also undergoes through the same psychoacoustic transformations as for the tonal speech. Based on the AMT and the perceptual noise estimation a generalized time-frequency spectral subtraction algorithm is implemented and evaluated by a speech recognition front-end.

Finally, Chapter 8 summarizes the results and new findings, including recommendations and some future research directions in this field.

We would like to acknowledge the Signal and Information Processing Lab, the Center for Intelligent Information Processing Systems, and the School of Electrical, Electronic and Computer Engineering

(EE&CE) for their invaluable support. Special thanks are due to the heads of school, Dr. G. Bundell, (2005) and Dr. Brett Nener (2008), and other staff members of the School of EE&CE, too many to name here, for their generous support during this research.

We would like to express thanks to other fellow researchers at SIP Lab, especially Aik Ming, Oscar Chan and Marco Kuhne with whom we had many enlightening and fruitful discussions within and outside the sphere of this work, and for the valuable suggestions they provided from time to time.

Finally, we would like to thank our family members, especially Selina and Sajid, who have stood by us during this time.

Serajul Haque   serajul@ee.uwa.edu.au
Roberto Togneri   roberto@ee.uwa.edu.au
Anthony Zaknich   tonko@ee.uwa.edu.au

Signal and Information Processing (SIP) Lab
Centre for Intelligent Information Processing Systems (CIIPS)
University of Western Australia.

List of Tables

2.1 Phonetic groups and the 47 basic phonemes in the English language
according to phonation. 24

2.2 Typical phoneme durations of the vowels and the consonants [1]. . . . 25

2.3 A review of research to-date on application of perceptual features for
speech recognition. 52

2.4 Comparison of research in Table 2.3 based on a HMM standard baseline
performance for speech recognition with perceptual features. 53

3.1 Consonants and the corresponding vowel tokens in the UCLA-SPAPL
CV corpus. 74

3.2 Measures of separability, J_1 and J_2, among the three vowel classes /a/,
/ee/ and /oo/ based on LDA of the within-class and between-class
scatter matrices for the base ZCPA and ZCPA with synaptic adaptation. 78

3.3 Recognition rates (%) of the ZCPA and ZCPA with synaptic adaptation
(ZCPA_ADP) for isolated digits (TIDIGITS corpus) with training and
testing with male utterances in four types of additive noise. 79

4.1 The two-tone suppression effect on the ZCPA interval histogram count.
It is observed that there is a reduction of histogram counts in the vicinity of the suppressor as the probe tone approaches the stronger suppressor tone fixed at 1 kHz. 93

4.2 Comparison of frequency discrimination coefficient, FD, of CV utterances between synaptic adaptation and two-tone suppression using the
ZCPA. 95

4.3   Continuous density HMM recognition rates (%) of the ZCPA with two-
      tone suppression compared with the base ZCPA and ZCPA with synap-
      tic adaptation for isolated digits (TIDIGITS corpus) with independent
      male speakers in four types of additive noise.   96

5.1   Equal loudness coefficients $\{L_e\}$ for 16 channels at ERB frequency scale
      for a frequency range of 10-3500 Hz.   102

5.2   Derivation of the compression angle and % compression from the com-
      pression coefficient, C.   110

5.3   Comparison of recognition rates (%) between the base MFCC_0 and
      MFCC_0 with equal loudness normalization and asymmetric compres-
      sion (MFCC_COMP), using male connected digits (TIDIGITS corpus)
      in clean condition and four types of additive noise.   112

6.1   Comparison of recognition rates (%) between the base ZCPA, and
      ZCPA with CN processing strategy, (ZCPA_AUD), with isolated digits
      (TIDIGITS corpus in four types of additive noise).   129

6.2   Data variance of MR output, $y_r(m, f)$ (Fig. 6.8) for the utterance
      'one' in white noise obtained with τ=0.1 ms and τ=10 ms at channel
      frequencies of 500 Hz and 2500 Hz. It is observed that the variance is
      reduced at a higher time constant for both high and low frequencies. .   132

6.3   Comparison of recognition rates (%) of ZCPA with MFCC_0, and ZCPA_AUD
      with MFCC_0 with delta features (26-dimensional MFC_DEL), in iso-
      lated TIDIGITS with male utterances in clean condition and in white
      and factory noise.   135

7.1   Percentage of T-F cells in perceptual noise which are masked by the
      masking threshold for an utterance 'one' in white noise.   148

7.2   Word Recognition rates (%) of conventional PLP processing compared
      with psychoacoustic spectral subtraction without and with the masking
      threshold for isolated digits and white additive noise.   154

xiv

Chapter 1

Introduction

The objective of this book is to investigate the role of human auditory perceptions in automatic speech recognition, and to determine the perceptual features which are relevant for speech recognition applications. Automatic speech recognition (ASR) is one of the most important research areas in the field of speech technology and research [1], [2]. It is also known as the recognition of speech by a machine or, by some artificial intelligence. Its main function is to transform an acoustic speech signal, captured by a microphone, a telephone, or other transducers, to a text sequence, usually in terms of a sequence of words. These may be used for applications such as document preparation and dictation, automated dialog systems, voice activated device controls, database access, hands-free applications as in car phones or voice-enabled PDAs, web enabling via voice, to name but a few. In short, it forms the basis for a man-machine interface for human-computer interactions.

The human vocal tract is represented by a time-varying filter excited by one or more sources. More generally, the vocal folds vibrate to generate a glottal wave, which acts as a resonator to modify the shape of spectra. Peaks of these acoustic spectra are referred to as formants. These correspond primarily to the steady-state vowels in speech with energies concentrated mostly in the low frequency regions. On the other hand, most consonants (except nasals and semi-vowels) are the high frequency components or brief transients with lower energies than the vowels. The transition region between a consonant and a vowel is dynamic. The primary aim of a speech recognition system is to identify the speech articulation cues robustly in presence of speaker variability, word perplexity, noise and channel conditions.

Research in speech technology in the past several decades has produced significant advances in ASR. The short-time analysis of speech dates back 50 years to the development of sound spectrograph [3]. Speech signals are pseudo-periodic with substantial random components, the statistical properties of which vary significantly over time. The short time Fourier transform (STFT) method of speech processing produces a series of feature vectors at a rate sufficient to capture the rapid instantaneous transitions in the spoken word within a short duration over which the statistical properties are reasonably stationary. This has contributed significantly to robust speech recognition. The linear prediction (LP) and autoregressive (AR) method of speech analysis utilizes the speech production process to determine the parameters of the vocal tract [4]. AR modelling yields the all-pole spectrum of the speech waveform, from which the frequencies and bandwidths of the individual resonances corresponding to the formant positions of the vocal tract may be obtained. Another significant improvement was the cepstral method of speech processing introduced by Bogert *et al.* [5] in 1963, which provided reduction of feature dimensionality and decorrelation of the vocal tract excitation and the glottal waveform. The recognition problem is one of determining the most probable word spoken, and usually done by template matching with a predetermined database of sound representation. The most popular and effective template matching is based on a set of hidden Markov models (HMM). It utilizes statistical modelling and a grammar rule to select the highest probability of some hidden or unobservable parameters of the speech from a sequence of observation feature vectors.

In spite of focused research in this field for the past several decades, the understanding of the acoustic-phonetic characteristics of speech, speech variability and speech perception is far from complete [6], and robust speech recognition with high reliability has not been achieved [7], [8]. In an ASR system, degradation may occur due to speaker variabilities. Another source of error may be the channel effects such as the convolutive noise added by the channel impulse response. The speech recognition process may work well in clean conditions but degrades significantly in noise conditions. The degradation is significant even in the presence of small additive and convolutive noise as a result of which the speech recognition may not function as a fully reliable system.

The ability of the human auditory system to recognize speech in adverse and noisy conditions has motivated speech researchers to include features of human perception in speech recognition systems, particularly in the early 1980s, when several computational models of the auditory periphery, based on physiological measurements of the response on individual fibres of the auditory nerve, were proposed [9-11]. These were generally referred to as "cochlear models". The primary objective of such models was speech processing for auditory research [12], hearing aids [13], cochlear implants [14], and to a lesser extent, for automatic speech recognition. These models of the peripheral auditory system usually include a perceptual filterbank corresponding to the critical bands of human hearing, a nonlinear rectification process, automatic gain control to compress the dynamic range that can be coded into the auditory nerve, synapse processing consisting of short-term synaptic adaptation, and lateral suppression. Some of these features enhance the speech segments, others suppress it to increase spectral contrasts or to suppress sources of stationary noise. It was shown that such approaches tend to provide robustness and recognition accuracy with a degree of improvement over conventional speech parameterizations either in clean conditions or in environment mismatch and noisy conditions [15], [16]. However, these improvements were achieved at a greater cost in computation time and storage requirements. Moreover, the use of perceptual features for speech recognition has resulted only in marginal improvements in many cases [17], [18], [19], and in specific cases even degraded ASR performance [20]. As a result, a decline in the interest in auditory models was observed until computing resources were able to meet the intensive computational requirements of such models. In recent years there has been a resurgence in perceptual processing of speech for speech recognition after some research have provided evidence that such processing may lead improved performance [19], [21], [7], [22], [27].

One of the objectives of this book is to identify the perceptual features which are relevant and useful for the improvement of ASR performance in clean and noisy conditions. Auditory perception is a psychoacoustic process in which the mapping between the acoustic speech and the perception is nonlinear. Equal acoustic changes often lead to widely varying perceptual effects. Due to the nonlinear characteristics, analytic treatment of auditory processing in the inner ear is often intractable, and relies substantially on experimental methods [1]. A distinction should be made between the two processes of psychoacoustic perception, which is a dynamic process, and speech recognition, which is primarily a static process. Auditory models for ASR integrate some of the the dynamic features of psychoacoustic perception with the static speech features. On the other hand, conventional speech processing for ASR usually ignores the detailed effects of perceptual features related to psychoacoustic perception. A comparison between a conventional and the perceptual speech recognition process is shown in Fig. 1.1. A second objective of this book is to include some of the effects of psychoacoustic perception into conventional speech processing for the purpose of enhancing automatic speech recognition performances.

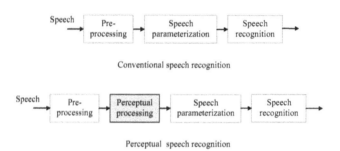

FIGURE 1.1: Perceptual speech recognition process compared with conventional speech recognition.

1.1 Methodology

Three methodologies are employed in this book. Firstly, many of the experimental investigations utilize temporal processing of speech using time domain filtering in the pre-processing stage. The temporal-place representation is much less affected by background noise than the rate-place representation [25]. On the other hand, spectral analysis is particularly effective in identifying the frequency components present in the speech, and for identifying the steady-state acoustic sounds. Auditory functions are usually dynamic with substantial transient properties. The auditory nerve is more responsive to changes than to steady inputs. Transients, unlike steady-state emissions, can not be fully characterized by the spectrum because, by their very nature, transients are time-varying, or non-stationary, signals [26], although short term Fourier transform processing may alleviate this to some extent. In many cases, the fine temporal structures and time variations are removed by the spectral feature extraction methods in the log spectral domain [27]. Recursive filtering which can simulate many of the psychophysical behaviors of human perception, is better implemented in the time domain than in the spectral domain because of the aliasing problem and the requirement of inverting an infinite dimensional Fourier transform. Time-varying cues are also useful in identifying CV (consonant followed by a vowel sound) transitions [28]. Typical speech contains about 12 phones/second. At faster rates, acoustic transitions between sounds become increasingly important. However, time domain processing is inherently slower, whereas the advantage of spectral processing is faster processing time.

In some of the implementations in this research, a zero-crossing algorithm for feature extraction from the input signal is utilized for speech parametrization. For this purpose, the zero-crossing peak amplitudes (ZCPA) [7] auditory model is selected. The ZCPA is based on the principle that any stimulus periodicity in the filter subband can be extracted from the zero-crossing intervals, the inverse of which shows up as a dominant frequency corresponding to the formant peaks within that subband. This is known as the dominant frequency principle [29]. The dominant frequency in a subband has more power than others, which makes the zero-crossing algorithms more robust in noisy environments. Moreover, the feature extraction process by zero-crossing analysis is computationally less complicated due to the fact that this analysis method is amenable to relatively simple transformations of the observed data. Other methods such as spectral analysis require a sophisticated transformation from the time domain to the frequency domain to obtain useable features from the measured data [30].

Secondly, the performances with the psychoacoustic effects were evaluated utilizing the continuous density HMM using clean and noisy speech, and measured in terms of the word recognition rates. psychoacoustic effects in speech processing and in speech enhancement are usually evaluated on human subjects by listening tests, such as the mean opinion score (MOS) and the dynamic rhyme test (DRT), nonsense syllable test, hearing-in-noise, etc., or by other perceptual measures for measuring the improvements in speech intelligibility. However, in this research the performances with the psychoacoustic effects were evaluated by the HMM method which performed digit recognition using the TIDIGITS corpus and the NOISEX 92 noise database. Both isolated digits and connected digits recognition were performed. Isolated digits recognition has been employed in cases where the main objective was to evaluate the relative performance benefits between two algorithms using the ZCPA auditory model. Connected word recognition was used to evaluate the ASR performance of the standard MFCC parametrization. However, isolated word recognition gives better recognition performances than the connected digit recognition. Statistical significance tests were performed to establish that the differences in error-rates between two algorithms tested on the same data set using isolated and connected word recognition are statistically significant.

Thirdly, for the feature extraction process, larger time windows and filters with appropriate time constants consistent with human perception were utilized. Timing in speech is often important perceptually. Since time domain processing is based on extracting spectral information from periodicity, it generally utilizes larger window lengths than the FFT methods. Auditory models do not make strong assumptions of quasi-stationarity. This is consistent with human auditory functions which have longer integration times with higher time constants. For example, synaptic adaptation and forward masking have a time constant varying from 40-80 ms [31]. However, STFT processing of speech using a filterbank approach (e.g. MFCC) is based on smaller stationary window lengths of 20-25 ms. Too short a window length produces a poor spectral representation of the speech, which usually degrades in noise condition, and is also sensitive to temporal effects of voicing [28]. Performance at higher SNR decreases with decreasing window lengths [28]. This has motivated several researchers to emphasize time interval processing with longer time constants for integration into the speech recognition systems [32], [33].

Chapter 2

Review of auditory perceptual features and speech recognition

2.1 The peripheral auditory system

Human speech perception is the aggregate responses to the various complex stimuli contained in an acoustic speech waveform. The function of the peripheral auditory system is to convert the incoming acoustic speech signals into perceptual representations in the form of an ensemble of neural signals [34], [12]. The peripheral auditory system is divided into three sections - the outer ear, the middle ear and the inner ear, as shown in Fig. 2.1. The outer ear consists of the pinna and the auditory canal which terminates in the eardrum or the tympanic membrane. The auditory canal has a length of about 2.5 cm and is an acoustic resonator which increases the ear's sensitivity to sounds in the higher frequency range of 3-20 kHz.

The middle ear is the link between the eardrum and the cochlea which is located in the inner ear. The cochlea, where speech encoding takes place, is a $2\frac{1}{2}$ turn snail shaped cavity filled with fluid. At one end of the tube, which is called the basal end, there are two flexible membranes called the oval window and the round window. The oval window connects the cochlea acoustically to the middle ear cavity

9

through three ossicles which are known as the malleus, the incus and the stapes. When the eardrum is stimulated by sound, the malleus attached to the outer side of the eardrum, transmits its vibration through the incus to the stapes. The stapes motion impinges on the oval window of the cochlea. When the oval window is pushed in by the stapes, the fluid in the cochlea is set in motion, which bulges back out through the round window. The cochlea is divided by a flexible cochlear partition called the basilar membrane (BM). It separates the tube lengthwise into two chambers called the ducts or the scalae. The BM is more rigid at the basal part of the cochlea and more flexible at the apical part of the cochlea. The fluid motion in the cochlea causes the BM to vibrate according to the frequency of the incoming acoustic signal.

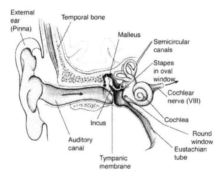

FIGURE 2.1: Structure of the peripheral auditory system consisting of the outer ear, the middle ear and the inner ear. The outer ear consists of the pinna and the auditory canal which terminates in the eardrum or the tympanic membrane. The middle ear is the link between the eardrum and the cochlea, which is located in the inner ear (Reproduced from Holmes, 1999).

About 3500 inner hair cells (IHC) are arrayed in a single row along the top of the BM partition in a structure known as the organ of Corti. Each IHC is connected to about 8-10 auditory nerve fibers, each having a different diameter. All the auditory information is carried by these approximately 28,000 nerve fibres that originate at the inner hair cells [35], [36]. Fine hairs called the stereocilia, protrude from the end of the inner hair cells. The mechanical vibrations of the BM cause bending movements of the stereocilia attached to the IHCs, which are transduced into neural firings of auditory nerve fibres [35]. The neural excitations are transferred to the auditory cortex as electrical impulses via the cochlear nucleus, which are composed of many cell groups, and the subsequent auditory pathways leading to the auditory cortex. It is suspected that there are complex feature detectors in these higher stages of the auditory system, but little is known of their functional behavior [7].

In addition to the IHC, there are three rows of about 12,000 outer hair cells (OHC) attached to the BM. The synapse is the junction between the hair-cell and the afferent nerve fibres. About ninety percent of the auditory nerve (AN) fibres synapse directly with the IHC, while the remainder synapse with OHCs. The OHCs do not send information about the sound to the brain. Instead, they function as part of an active amplifier and automatic gain control (AGC) [35]. The OHCs are responsible for our acute sensitivity and superb frequency-resolving capabilities. These also control the various nonlinear phenomena that can be observed in the mechanical response of the BM and the neural response of the auditory nerves. One of the primary functions of the peripheral auditory system is to combine high sensitivity with a large dynamic range. This is primarily achieved by the outer hair cells, which amplify weaker signals so that they are audible, and compress sound with higher intensity so that these are within tolerable hearing levels [35], [36].

Beyond the peripheral auditory system, the auditory nerves enter the cochlear nucleus, which has at least three major divisions on the basis of morphology [37]. The auditory nerve bifurcates into an

11

ascending branch that innervates the anteroventral cochlear nucleus (AVCN) and a descending branch that innervates both the postventral cochlear nucleus (PVCN) and the dorsal cochlear nucleus (DCN) [37]. Each of the divisions contains a complete representation of the audible frequency range. At least 22 different types of neurons have been anatomically distinguished [38]. Based on patterns of response to short tone bursts, different response types have been distinguished which include primary-like, chopper, onset, buildup and pauser, etc. [38], [37].

2.2 Basilar membrane transduction and the critical bands

The pioneering work of von Békésy [39] established that the cochlea is capable of analyzing the input acoustic signal into different frequencies due to the vibration of the BM to an input acoustic signal. The BM, about 32 mm in length, responds to sound vibrations in the incoming speech and human ears hear the frequencies most sharply at which the BM responds or vibrates the most. It produces a tonotopic encoding of the stimulus, that is, the signal is decomposed into its component frequencies that generate a place code in the pattern of the AN fibre activities. These frequencies are referred to as the characteristic frequencies (CF) and are given by the Greenwood [40] formula,

$$F = A(10^{ax} - 1) \tag{2.1}$$

where F is the frequency in Hz, x is the normalized distance along the BM (with values between 0 and 1), A=165.4 and a=2.1 are appropriate constants for humans. Thus the BM acts like a tuned bandpass filterbank, where the tuning frequency is a function of the distance x along the BM.

The human auditory system has a limited frequency resolution.

The perceptually uniform measure of frequency can be expressed in terms of the width of the "critical bands" [41]. These bandwidths correspond roughly to 1.36-mm spacings along the BM. This indicates that the 32-mm length of the BM may be simulated by 24 bandpass filters. The critical bands and the Bark frequency scale [41] is shown in Appendix C.2. The bandwidth is less than 100 Hz at the lowest audible frequencies, and more than 4 kHz at the high end. Altogether, the frequency range 20-15500 Hz can be partitioned into 24 critical bands based on classical masking experiments involving a narrowband masker and probe tone. A new unit for frequency termed "Bark" (after Barkhausen) is used, where one Bark is taken as the width of one critical band [41]. For frequencies $<$500 Hz, it converts linearly to (freq/100) Bark. For frequencies $\geq$500, it is given by

$$9 + 4\log_2(\text{freq}/1000) \quad \text{Bark}. \tag{2.2}$$

Another perceptual frequency scale based on psychophysical experiments of human perception is the equivalent rectangular bandwidth (ERB) rate, proposed by Moore and Glasberg [42]. Moore and Glasberg [42] have revised Zwicker's [41] loudness model to better explain how equal-loudness contours change as a function of level and why loudness remains constant as the bandwidth of a fixed-intensity sound increases up to the critical bandwidth. The ERB of the auditory filter is assumed to be closely related to the critical bandwidth. But the bandwidths for CB are constant below 500 Hz, whereas for ERB the bandwidths continue to decrease below 500 Hz. ERB is measured using the notched-noise method rather than the classical masking experiments involving a narrowband masker and probe tone for CB. As a result, the ERB is said not to be affected by the detection of beats or intermodulation products between the signal and masker. Since this scale is defined analytically, it is also more smoothly behaved than the

Bark scale data. The ERB-rate is given by

$$\text{ERB-rate} = 11.17 \log_e \left(\frac{f + 0.312}{f + 14.675} \right) + 43.0 \qquad (2.3)$$

where f is the frequency in kHz.

The mechanical vibrations of the BM cause bending movements of the stereocilia attached to the IHCs which causes ionic channels in the cell membrane to open and release neurotransmitters. All cells are inside a membrane that can permit charged ions to flow from inside to outside of the cell and vice versa. Control of this chemical flow is through molecules that are embedded in the cell membrane. Many of the chemicals both inside and outside the cell are charged so that this molecular flow can cause changes in the voltage potentials of the cells. The flow of potassium and sodium ions create voltages needed to trigger neural spikes (action potentials or the neural discharge rate) in the nerve fibres. The detailed mechanism of spike generation in the auditory nerve is explained by the Hodgkin-Huxley model of neural firing [43].

In terms of auditory processing, the neural transduction from the input sound stimulus taking place in the inner ear as action potentials is explained by two models: the hair cell model and the synapse model [44], [45]. This is shown in Fig. 2.2. The input to the IHC model is the motion of the BM vibrations, and the output is a stream of neurotransmitter materials releasing events corresponding to the input signal [45]. The synapses from the inner hair cells to the auditory nerve (AN) fibres is known as the synapse model [1]. The output of the synapse model is an array (ensemble) of instantaneous and probabilistic neural firings (discharge) of the auditory nerve fibres which propagates to the higher auditory system. The inner hair cell model and the synapse model as a whole is referred to as the IHC-Synapse model, or the cochlear model. In a peripheral auditory model, speech features are extracted from the spectral shape or periodicities of the AN excitation patterns. Two parallel theories are used to represent the coding of the acoustic signals in the discharge rate of an AN fibre. These are the rate-place theory and the temporal-place theory [7], [46].

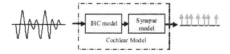

FIGURE 2.2: The IHC-synapse model (also known as the cochlear model).

2.2.1 Rate-place theory

The rate-place theory assumes that each ascending fiber innervating an inner hair cell responds best to a frequency, known as the characteristic frequency, which is determined based on specific location of the hair cells on the BM. A single auditory nerve conveys the spectral contents of the stimulus by an average neural discharge of the fibre [46]. One of the flaws of this model is that it does not take into account the saturation effect at higher sound-pressure levels [34] which causes a loss of definition in the spectral pattern of the vowels, yet the auditory system does not lose the ability to discriminate articulatory cues at these higher intensity levels. Several researchers have developed the synchrony (temporal) model to alleviate this problem.

2.2.2 Temporal-place theory

This theory takes into account the timing information associated with the input stimulus. The time interval between neural discharge of the auditory nerve fibres is compared with the reciprocal of the fibre's CF.

It is found that the neural discharge of an auditory nerve fibre occurs at a particular phase of the input stimulus and the auditory nerve fiber tends to discharge in synchrony to it. This synchronous neural representation contains the useful frequency information in the input signal [46]. The auditory nerve fibers are capable of locking or synchronizing to harmonics of the stimuli that correspond to the formant frequencies [47]. However, at higher frequencies, the phase locking capabilities are greatly reduced and there is a loss of synchrony observed in the auditory nerve response. The temporal-place representation is much less affected by background noise than the rate-place representation [25]. Synchronized discharge rate averaged over several fibres with CFs near ± 0.25 octave of a stimulus component may be used as a measure of the population temporal response to that component, and is referred to as the average localized synchronized rate (ALSR) [25].

2.3 Psychoacoustics

Psychoacoustics is the study of the subjective human perception of the objective properties of sound (e.g., frequency and intensity) and is the process of correlation of psychoacoustic phenomena with physiological measurements [2]. Each of the objective properties of sound may be related with a corresponding perceptual variable. For example, the perceptual variable of frequency is called the pitch, the perceptual variable of intensity is called the loudness, and the perceptual variable of spectral shape is called the timbre.

The transduction of an input acoustic signal to neural representation in the auditory nerves is accomplished through a series of psychoacoustic transformations. These transformations are usually nonlinear such that an increase in the input stimulus does not produce a corresponding increase in the output [2]. As a result, the output cannot be determined by simply superimposing the inputs. Most of the perceptual features due to the psychoacoustic transformations increase in a logarithmic scale, which has motivated the use of the decibel (dB) scale in psychoacoustic analysis. The dB scale is defined as the logarithmic unit of measurement that expresses the magnitude of a physical quantity such as the speech intensity relative to a specified reference level. Besides the nonlinear frequency response of the BM, several other multistage nonlinear transformations take place in the inner ear, such as dynamic range compression (compressive nonlinearity), synaptic adaptation and masking, and the two-tone suppression. Some of the essential perceptual features of human auditory system are stated below.

2.3.1 Equal loudness curves of human perception

Two important acoustic parameters are sound pressure (P) and sound intensity (I). Sound pressure P is the sound field value whereas sound intensity I is a sound energy value $(I = P^2)$. The unit of sound intensity is watts/m^2 and is measured by sound intensity level (SIL) which is given in dB by

$$L_I = 10 \log_{10} \frac{I}{I_0} \qquad (2.4)$$

where $I_0 = 10^{-12}$ is the intensity level corresponding to 0 dB SIL and is taken as the reference level [48]. The unit of sound pressure is Pascal (Pa) and it is measured by sound pressure level (SPL) which is given in dB by

$$L_P = 20 \log_{10} \frac{p}{p_0} \qquad (2.5)$$

where $p_0 = 2 \times 10^{-5}$ Pa is the sound pressure level corresponding to 0 dB SPL and is taken as the reference level [48].

An objective measure of loudness is sones (L), which is defined as the loudness of a tone relative to a 1000 Hz tone at 40-dB SPL. The loudness level is usually measured in phons, LL, which is defined to be numerically equal to the intensity level, L_I in dB, at a frequency of 1000 Hz [49]. The relationship between the sones and the phons may be expressed as

$$\log_{10}(L) = 0.033(LL - 40)$$
$$= 0.033LL - 1.32 \qquad (2.6)$$

Using the definition of the loudness level, LL, and using Eqn. (2.4) in Eqn. (2.6)

$$\log_{10}(L) = 0.033(10\log_{10}(I) + 120) - 1.32$$
$$= 0.33\log_{10}(I) + 2.64. \qquad (2.7)$$

which reduces to the intensity-loudness relationship, or the power law of hearing [49]

$$L = 445I^{0.33}. \qquad (2.8)$$

Fig. 2.3 shows the equal loudness curves (Fletcher-Munson curves) of human perception to a tone frequency as a function of the loudness level in phons. The curves were obtained from psychophysical experiments where the listener adjusted the intensity at a given frequency as the frequency was varied until it was judged to be of equal intensity to a standard 1000-Hz tone. It is seen from these curves that the ear is most sensitive to sound at approximately 3.8 kHz and that the sensitivity decreases at higher frequencies.

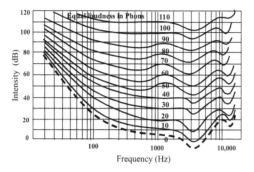

FIGURE 2.3: The standardized equal loudness set of curves of human perception (Fletcher-Munson loudness curves) as a function of frequency and the loudness level phons. The phons, by definition, is equal to the intensity level in dB at 1 kHz (adopted from [50]).

2.3.2 Half-wave rectification

Due to the vibration of the acoustic signal, the fluid in the cochlea travels back and forth and causes the cilia attached to the IHC to bend. When the cilia are bent one way, the hair cell stimulates the primary auditory neurons to fire. When the cilia are bent the other way, no spikes are generated. However, there may be some release of neurotransmitter, which is not sufficient to generate spikes. Thus, the nerve fibers translate the motions of the BM into a sequence of spikes, which tend to be initiated on only one half-cycle of the BM movement. Hence, the inner hair cells acts as half-wave rectifiers for the velocity of the motion of the fluid. For this reason, conventional auditory models provide some means of half-wave rectification in the transduction process occurring in the BM [9].

2.3.3 Compressive nonlinearity

One of the important perceptual properties of the peripheral auditory system is that the transduction of basilar membrane (BM) vibration into auditory nerve (AN) discharge is nonlinear. The neural firings

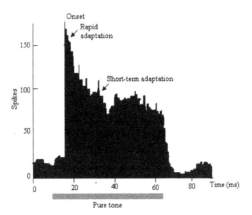

FIGURE 2.4: Poststimulus time histogram (PSTH) of an auditory nerve fibre to a pure tone burst showing the rapid and short-term synaptic adaptation (adopted from [35]).

(discharge) of auditory nerve fibres depend on the strength of the auditory input more than on the frequency. As the sound pressure level is increased, there is a corresponding increase in the neural discharge rate [34]. At higher SPL levels, there is a saturation effect and any further increase in the stimulus intensity will not produce a corresponding increase in the neural discharge rate. This is referred to as the compressive nonlinearity [12], [51]. The mechanical properties of the inner ear are linear at levels below 20 dB SPL, but compressive between 20-100 dB SPL [52]. The compression is higher for a higher CF than for a lower CF [53].

2.3.4 Synaptic adaptation

In response to tone bursts, a single auditory-nerve fibre exhibits an increased neural discharge in the initial 15 ms corresponding to the signal onset. This decays monotonically in time, reaching a steady-level within about 50 ms. This decrease in response rate, referred to as "synaptic adaptation", has been determined by physiological experiments in observed responses to pure tones [54], [31]. The decay consists of an initial rapid phase with a time constant of 3 ms (rapid synaptic adaptation). It is followed by a slower exponential decay with a time constant of about 40 ms (short-term synaptic adaptation). Fig. 2.4 shows a post-stimulus time histogram (PSTH) for an afferent nerve fibre. It is observed from the figure that in the absence of the stimulus the auditory nerve fibre produces spikes at a small but discernable rate. This is called the spontaneous rate. When the tone is removed, the spike rate decreases to slightly lower than the spontaneous rate for a short time before resuming its normal spontaneous rate. The synaptic adaptation process indicates that the auditory nerve is more responsive to changes than to steady inputs. The rate of fall from the onset transient down to the synaptic adaptation level is independent of the intensity of the tone signal level [54].

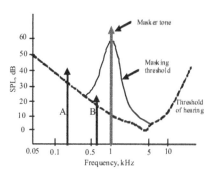

FIGURE 2.5: Simultaneous (frequency) masking in presence of a 1 kHz tone masker. Tone 'A' is not masked, whereas tone 'B' is masked by the masker.

2.3.5 Auditory masking

Temporal masking is another observed psychoacoustic phenomenon by which one sound (maskee) is made inaudible by the presence of another sound (masker), preceding it in time. This is known as forward masking. Masking may also be backward in which case the masker follows the maskee signal in time. Forward masking is believed to be a consequence of short-term synaptic adaptation [55]. Masking may also be simultaneous (frequency masking) in which case one sound is made inaudible by the presence of a louder sound at a neighboring frequency. It is also observed that a tone more easily masks a tone of higher frequency than one of lower frequency [57]. Fig. 2.5 shows simultaneous masking in the presence of a 1 kHz masker tone. Tone 'A' is not masked whereas tone 'B' is masked by the masker. A tone may be masked by noise. On the other hand, the noise may also be masked by a tone. In other words, "hiding the noise" underneath signal spectrum (that is, making the noise inaudible) is feasible because of human auditory masking [57], [56].

2.3.6 Two-tone suppression

Two-tone suppression is a phenomenon of the auditory nerve responses in which the neural firing rate in the region most sensitive to a probe (supppressee) tone is reduced by the addition of a second (suppressor) tone at a frequency near the probe tone [58]. Two-tone suppression arises due to wide-band nonlinearities in the BM mechanics [58], [59]. It determines how well auditory nerve fibres synchronize to specific formants at different stimulus intensities.

The difference between auditory masking and two-tone suppression is that in auditory masking a weaker sound is made inaudible by a stronger sound, whereas in two-tone suppression there is a reduction in magnitude of a weaker tone due to the presence of another stronger tone at a frequency higher or lower than the stronger tone.

Fig. 2.6 shows two-tone suppression in presence of a probe tone. The shaded region is the suppression region, enclosed by white squares.

The excitation region of the probe tone is enclosed by dark circles [60]. In presence of a probe tone within the excitation region, the presence of a second tone having any combination of frequency and amplitude within the shaded (suppression) region will cause a net reduction in the neural discharge of auditory nerve fibres.

Another suppression phenomenon, probably related to two-tone suppression, is the lateral inhibition, by which an auditory nerve reduces its own gain as well as the gain of others nearby by lateral distribution of its outputs to inhibitory synapses on neighboring auditory nerves [121]. This phenomenon is similar to the lateral inhibition that is present in human vision [63].

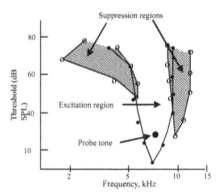

FIGURE 2.6: Two-tone suppression and excitation regions in presence of a probe tone. The shaded region is the suppression region, enclosed by white squares. The excitation region is enclosed by dark circles [60].

2.4 Articulatory phonetics

In linguistics, the fundamental unit of speech is the phoneme which is the minimal unit to distinguish between meanings of words, whereas a phone is a smallest identifiable unit found in a stream of speech [64], [1], [65]. Phonemes are composed of a family of similar sound groups. There are about 61 phoneme-like units in the TIMIT American English speech corpus [23]. Collapsing the allophones (variants of a particular phoneme) into the similar phones, 47 distinct phones can be identified. Depending on the vocal tract articulations and the mode of phonation, these can be divided into seven categories as shown in Table 2.1.

TABLE 2.1: Phonetic groups and the 47 basic phonemes in the English language according to phonation.

Phonetic groups	Sample phonemes
Semi-vowel and glides	l, r, w, y
Vowels	a, ae, ah, ao, aw, ax, axr, ay, eh, el
	em, en, er, ey, ih, iy, ow, oy, uh, uw
Stops	b, d, g, p, t, k
Affricates	jh, ch
Fricatives	s, Z, zh, nx, dh, dx, sh, f, hh, V, th
Nasals	m, n, ng
Silence	h#

The phonemes differ in their manner and place of articulation, that is, the state of the vocal cords vibrating or not vibrating (corresponding to voiced and unvoiced speech), and the degree and the location of the vocal cord constriction. The vowels are the largest class of phonemes in which air passes unrestricted through the oral cavities. On the other hand, stops (or plosives) employ a complete closure followed by a release (burst). Fricatives resemble stops, but instead of a total closure, maintain a narrow vocal tract constriction [64]. An affricate is a stop followed by a fricative. Whisper-like sound without vocal cord vibration due to large airflow through a constricted vocal tract is called aspiration. The main difference between aspiration and frication is that aspiration is broadband, whereas frication is high frequency only [1].

Most vowels are steady-state components of speech, represented by the the formant peaks of the vocal tract. Consonants are brief transients (voiced and unvoiced) but some include quasi-steady components which are also the dominant characteristics of vowels [64]. The onset/offset of speech sounds is inherently transient broadband events and the consonant-vowel (CV) transition is dynamic, that is, varies with time. It is generally recognized that the voicing and the steady state vowel sounds are largely low frequency with higher energies, and that the consonants are dominated by higher frequencies with low energies. The region below 500 Hz represents voicing and occasionally some first-vowel formant information, whereas most of the information-bearing spectrum of vowels and nearly all consonant spectral power lies above approximately 500 Hz [1].

Phoneme durations are important cues for ASR [1]. Determination of correct phone boundaries are critical for phoneme recognition. Phoneme duration varies with the speaking rate. Typical speech contains about 12 phones/s, but rates as high as 50 phones/s are possible [1]. This makes determination of phone boundaries more difficult. The average duration of a phone is about 80 ms [1]. Table 2.2 shows the typical phoneme durations of the vowels and the consonants.

TABLE 2.2: Typical phoneme durations of the vowels and the consonants [1].

Phoneme groups	Durations (ms)
Vowels and dipthongs	80-150
Consonants	40-80

2.5 Computational models for speech processing

The fundamental task of ASR is to transform human speech to a form which can be interpreted by a computer, a control device or an interactive network. A speech recognition system consists of three broad processing steps. The first step is the pre-processing which converts the incoming speech to acoustic features. The second step is the speech parametrization which converts the acoustic features to speech vectors suitable for processing by a speech recognition system. The acoustic pre-processing and parametrization are usually accomplished by an ASR front-end. The next step is the speech recognition process, the function of which is to model the acoustic feature vectors with sufficient details and accuracy and to identify correctly and robustly the original spoken words. This stage is known as acoustic and language modelling and the decoding process itself.

A computational model of the speech process is the representation of the phonological (symbolic) and phonetic (numeric) aspects of human speech, which can be subject to discrete time computation for computer speech recognition and understanding, synthesis, analysis, enhancement, coding, speaker identification, and so on [1].

An auditory model usually integrates psychoacoustics and physiology of human speech perception in a computational model. Modeling the peripheral auditory system is complicated by the fact that little is known about the exact mechanisms of the auditory periphery for detailed construction of the model. In most cases, analytic treatments are intractable because of multistage nonlinear transformations. As a result, most auditory models rely heavily on experiments [7]. Analytic or mathematical representations of such models are difficult, and whenever possible the parameters for such model are experimentally determined. A particular output cannot be predicted by a mathematical formulation and the interaction of its parameters. Rather, the outputs are obtained by the development of an algorithm from its known properties and behaviour. Experiments may take the form of a computer run, or subjective manual tests. However, there have been some efforts to analyze auditory models [15], [66], particularly by means of comparing or predicting an output for a given set of input variables, with respect to some reference performances. The disadvantage is that these are more computationally expensive than conventional speech parameterizations for ASR.

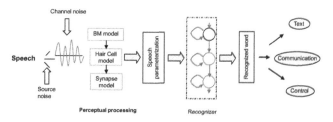

FIGURE 2.7: A computational auditory model for ASR system utilizing perceptual processing.

A computational auditory model for automatic speech recognition system as a front-end preprocessor is shown in Fig. 2.7. The perceptual processing consists of three main parts: the BM model, the hair cell model and the synapse model, as stated in Sec. 2.2.

2.6 Acoustic pre-processing and speech parametrization

For automatic speech recognition, the analog speech signal is converted to digital form by a sampling process. For telephone speech, the sampling rate is usually 6.4 kHz and for broadband speech, it is 16-20 kHz. Raw speech is first processed according to the functionalities of the outer ear, the middle ear and the inner ear in the pre-processing stage. Next, the acoustic features are parameterized as feature vectors to be able to be understood by a speech recognition system.

Each speech sound has particular spectral characteristics which change slowly in time, such as the steady-state vowels and the consonant-vowel transitions. However, transients and plosives are not slowly varying, but these constitute a smaller segment of a word utterance. Since the overall goal in pattern classification is to distinguish between examples of different classes, the goal of feature extraction (preprocessing and parametrization) is to reduce the variabilities within the same class, while increasing the variabilities between features that belong to different classes. The feature vectors should emphasize perceptually important speaker-independent features of the signal, and de-emphasize speaker-dependent characteristics. Additionally, it should be robust to acoustic variations but sensitive to linguistic content. Speech vectors are usually normalized to eliminate irrelevant factors in speech. Ideally, the feature space should be low-dimensional, because lower dimensionality means a smaller covariance matrix for the word utterance, which increases computational efficiency

2.6.1 Static parameters

Feature extraction may be performed in the frequency domain or in the time domain. Frequency or spectral domain processing requires using the spectrum of the speech signal in the form of discrete Fourier transform (DFT). The properties of the speech signal change relatively slowly with time. However, purely spectral analysis using the DFT is inadequate to represent the shift of spectral characteristics with time. Hence, time information is combined with the DFT by taking short time segments (usually 20 ms) durations over which the properties are relatively stationary, with one analysis performed every 10 ms. This is the short-time Fourier transform (STFT) processing method of speech. If $s(n)$ is the speech signal, and $w(n)$ is a window function, the formula for short-time energy is

$$E(n) = \sum_{m=0}^{N-1}[s(m)w(n-m)]^2 \qquad (2.9)$$

where $w(n-m)$ is a windowing function and N is the duration of the window. The energy function can also be used to locate approximately the time at which voiced speech becomes unvoiced, and vice versa, and to distinguish speech from silence. If the time window, $w(n)$, is of short duration, then its Fourier transform will be of a wide bandwidth which gives poor frequency localization. For any specific window type, its duration varies inversely with spectral bandwidth, i.e., the usual compromise between time and frequency resolution [1]. If the analysis windows are not chosen to be synchronous with any acoustic landmark, that is, not time aligned, the resultant features will be smeared over transition regions.

Fig. 2.8 shows the STFT analysis of speech using a 25 ms Hamming window for the CV utterance /ba/. The top panel shows the first 50 ms containing the bursts after the stop, the spectrum and the time-frequency plot (spectrogram) illustrating the high frequency components and transients in the consonant /b/. The lower panel shows the same information for the segment 310-350 ms containing the steady state portions of the vowel /a/. It is seen from the spectrograms that for the consonants (with the exception of nasals and semi-vowels) and transient portions, the energy is concentrated mostly in the high frequency regions, while for the steady state vowels, the energy is concentrated more in the low frequency regions.

There are several time domain methods for processing of speech. One such method is based on the short-term average zero-crossing rate. In the context of a discrete time signal $s(n)$, a zero-crossing occurs if successive samples have different algebraic signs,

$$s(n) = 0.5[\text{sign}(s(n)) - \text{sign}(s(n-1))] \qquad (2.10)$$

The algebraic sign of $s(n)$ is

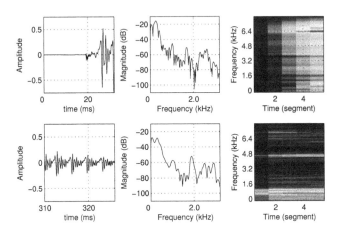

FIGURE 2.8: Short-time Fourier transform (STFT) spectrum and the time-frequency plot (spectrogram) analysis of speech of the CV utterance /ba/. The top panel shows the first 50 ms containing the bursts after the stop, illustrating the high frequency components in the consonant /b/. The lower panel shows the segment 310-350 ms containing the steady state portions of the vowel /a/. (In gray scale, darker shade implies lower intensity and lighter shade indicates higher intensity).

$$\text{sign}(s(n)) = \begin{cases} 1 & \text{for} \quad s(n) \geq 0 \\ -1 & \text{otherwise.} \end{cases}$$

This measure gives a reasonable way to estimate the frequency of a sine wave and useful spectral information can be extracted from the zero-crossing representation. A reasonable generalization is that if the zero-crossing rate is high, the speech signal is unvoiced, while if the zero-crossings rate is low, the speech signal is voiced [1].

Another time domain method for processing of speech is the autocorrelation function which may be used for estimating the pitch by detecting the periodicity in a signal. The formula for short-term autocorrelation as given in [1] is

$$R_n(k) = \sum_{m=0}^{N-1-k} s(m)w(n-m)s(m-k)w(n-m+k). \qquad (2.11)$$

$R_n(k)$ measures the similarity of a windowed signal $s(n)$ with a delayed version of itself, as a function of the time delay k, and has a maximum value at $k=0$.

Cepstral coefficients [5] are the most common representation of the spectral characteristics for speech recognition, which provide good decorrelation of the glottal excitation $e(n)$ from the speech envelope $v(n)$. The spectrum of the observed speech $s(n)$ is given by $|S(\omega)| = |E(\omega)||V(\omega)|$. After cepstral transformation, the first four or five transformed lower dimensional vectors are sufficient to preserve nearly all the information of the original vectors. Therefore, only the first few coefficients may be retained, allowing dimensionality reduction of the feature vectors. The n-th cepstral coefficient is given by

$$c(n) = \frac{1}{2\pi} \int_{-\pi}^{\pi} \log |S(\omega)| e^{j\omega n} d\omega \qquad (2.12)$$

where $|S(\omega)|$ is the spectrum of the observed speech. The separation of the vocal tract shaping, which primarily is the formant resonances of the vocal tract, from the glottal excitation may be implemented by linear filtering in the cepstral domain. Such a filtering in the cepstral domain is known as liftering.

The principal component analysis (PCA) technique provides a more powerful decorrelation which can greatly remove the linear dependencies between the sets of variables. Data reduction techniques such as PCA have been shown to be successful in speech recognition tasks, especially when speech is degraded by noise or spectral tilt [21], [15].

There are two broad approaches for speech parameterization : a) linear prediction, and b) filterbank analysis. In linear prediction (LP) analysis [4], the vocal tract transfer function is modeled by an all-pole filter with the transfer function

$$H(z) = \frac{1}{\sum_{i=0}^{p} a_i z^i} \qquad (2.13)$$

where p is the number of poles and $a_0=1$. The filter coefficients a_i are chosen to minimize the mean square filter prediction error summed over the derivative window. Usually the autocorrelation method is used to perform this optimization.

The human ear resolves frequencies nonlinearly across the auditory spectrum. Experimental results suggest that designing a front-end to operate in a similar nonlinear manner may improve recognition performance. A popular alternative to linear prediction is, therefore, the filterbank analysis which forms the basis of the mel-frequency cepstral coefficients (MFCC) parametrization [67]. The filterbank amplitudes are highly correlated and hence, a cepstral transformation is performed if the data is to be used in a HMM based recognizer with diagonal covariances [68], although other diagonalisation approaches are also possible. The filters used are triangular and are equally spaced along the mel-scale which is defined in [68] as

$$\text{mel}(f) = 2595 \log_{10}\left(1 + \frac{f}{700}\right). \qquad (2.14)$$

To implement the filterbank, the window of speech data is transformed using a fast Fourier transform and the magnitude (or magnitude squared if power spectrum is desired) is computed. The magnitude coefficients are then binned by multiplying them with each triangular filter. Each bin holds a weighted sum representing the spectral magnitude in that filter channel, as defined by

$$S_i = \sum_f |X(f)|H_i(\text{mel}(f)) \qquad (2.15)$$

where $X(f)$ is the output of the FFT at frequency f, f ranges over the set of frequencies evaluated in the FFT, and $H_i(\text{mel}(f))$ is the height of the i-th bin at the frequency $\text{mel}(f)$. A log of the binned magnitude is taken and the discrete cosine transform (DCT) is performed for decorrelation to obtain the cepstral coefficients c_j [69],

$$c_j = \sqrt{\frac{2}{N}} \sum_{i=1}^{N} \log(S_i) \cos\left(\frac{\pi j}{N}(i - 0.5)\right) \qquad (2.16)$$

where N is the number of filterbank channels and $j = 0, \ldots, J - 1$, J being the number of cepstral coefficients.

2.6.2   Perceptual linear prediction (PLP)

An enhancement to the LPC analysis is perceptual linear prediction (PLP) [70] which integrates three properties of human perception to obtain the audible spectrum. These are critical band filtering, equal loudness pre-emphasis and cube root power law of hearing (intensity-loudness conversion) which renders this method more robust in speaker-independent condition. The power spectrum is calculated using a FFT and computing its squared magnitude. This is filtered with overlapping critical band filters spaced roughly at 1-Bark intervals. The equal loudness pre-emphasis which approximates the human audibility sensitivity to react at different frequencies at about 40 dB (Fig. 2.3) is given in [70] as

$$E(\omega) = \frac{(\omega^2 + 56.810^6)\omega^4}{(\omega^2 + 6.310^6)^2(\omega^2 + 0.3810^9)}. \qquad (2.17)$$

The spectral amplitude is then compressed following the power law of hearing (instead of a conventional log compression), using the Eqn. (2.8). Aside from matching this property of human hearing, the effect of this step is to reduce amplitude variations for the spectral resonances [2]. An inverse DFT is performed to obtain something like autocorrelation coefficients (since a log has not been computed), retaining only the cosine components. A spectral smoothing is performed to reduce the effects of nonlinguistic sources of variance of speech signals. This is accomplished by constructing an all-pole autoregressive model using Eqn. (2.13) and solving it by the Levinson-Durbin recursion. The coefficients are then converted by cepstral transformation. The PLP processing of speech is shown in Fig. 2.9.

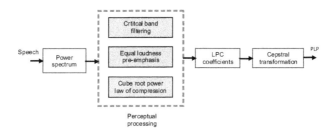

FIGURE 2.9: Perceptual linear prediction (PLP) processing of speech.

2.6.3 Temporal filtering and dynamic parameters

Temporal processing method of speech are commonly used to aid in performance and robustness in ASR. Usually, the spectral or spectrally related values, such as cepstra, are processed in the time domain to enhance or preserve the speech carrying modulations or to add speech dynamics information to the speech vectors. One such processing is RASTA [71] filtering which takes advantage of the fact that rate of change of temporal properties of nonlinguistic components such as noise and channel distortions lie outside the typical rate of change of the vocal tract shape. RASTA processing suppresses the spectral components that change more slowly and quickly than the typical range of speech.

The performance of the speech recognition system can be greatly enhanced by adding time derivatives (the rate of change over time) to the basic static parameters [72]. The first order regression coefficients (delta coefficients) are computed using the following regression formula given in [68] as

$$d_t = \frac{\sum_{\theta=1}^{\Theta} \theta(c_{t+\theta} - c_{t-\theta})}{2\sum_{\theta=1}^{\Theta} \theta^2} \tag{2.18}$$

where d_t is the delta coefficient at time t computed in terms of the corresponding static coefficients $c_{t+\theta}$ and $c_{t-\theta}$ and Θ is the number of frames taken in the delta computation before and after the centered frame. The second order regression coefficients (delta-delta) are computed using Eqn. (2.18) a second time on the delta coefficients. For a regression over 3 frames to the left and right of the centered frame, $\Theta=3$.

A convolution effect in the time domain (e.g., caused by passing the signal through a linear filter or a communication channel) corresponds to a multiplication in the frequency domain, which is equivalent to a sum in the log power domain. Therefore, a very simple and effective technique of improving robustness in the presence of convolutive disturbances is cepstral mean subtraction (CMS), by which the irrelevant and unwanted dc or mean value of the log spectrum is subtracted from the speech in the cepstral domain.

2.7 Automatic speech recognition

In ASR, the recognition process is to correctly and robustly identify the spoken word by a recognizer system from the feature representation of the spoken word. The problem of speech recognition is a difficult one because of several reasons. Firstly, natural speech is continuous without pauses, which makes it difficult to ascertain the word boundaries. Secondly, natural speech is affected by the differences in rates of speech, pronunciation, mood, style and age across speakers. Thirdly, if the vocabularies (lexicon size) are large, recognition becomes more difficult because there are more confusable words. The same words may have different meanings depending on the context. Fourthly, recorded speech is variable over room acoustics, microphone and channel characteristics and noise conditions. Even a simple change of microphone may degrade recognition performances. Additive broadband noise causes few voicing errors, but many place errors. Noise with a flat spectral level (white noise) tends to degrade frequencies with low energies, that is, mostly higher frequency components. A human may discriminate these differences to maintain robustness, where a machine recognition system cannot. Therefore, a recent trend in auditory research is aimed at accounting for human auditory performances, using the results of physiological auditory modelling and statistical decision/detection theory [7], [16], [73], [19].

2.7.1 Deterministic speech recognition-pattern classifiers

In the deterministic speech recognition based on template matching, some similarity measure between a speech segment and a reference speech segment is established. This reference can be a prototype of the same features that are extracted from speech during the training process of a representative number of examples of each class. A deterministic classifier, such as a minimum distance classifier, utilizes a metric distance (e.g. Euclidean) between the input features and the reference models or the prototypes to find the closest match [74]. For a simple minimum distance classifier with j prototypes, z_i, where $0 \leq i < j$ for each input vector $\mathbf{x}$, i is selected such that it minimizes the distance D_i,

$$\underset{i}{\mathrm{argmin}} \, [D_i = \sqrt{(\mathbf{x} - \mathbf{z}_i)^T (\mathbf{x} - \mathbf{z}_i)}]. \tag{2.19}$$

Another important distance measure is the Mahalanobis distance. If $\boldsymbol{\mu}_x$ and $\boldsymbol{\mu}_z$ denote the mean values of the vectors $\mathbf{x}$ and $\mathbf{z}$, then the Mahalanobis distance from $\mathbf{x}$ to $\mathbf{z}$, denoted by d is given by

$$d^2 = (\mathbf{x} - \mathbf{z})^T \boldsymbol{\Sigma}^{-1} (\mathbf{x} - \mathbf{z}) \tag{2.20}$$

where $\boldsymbol{\Sigma}$ is the covariance matrix. The Mahalanobis distance takes into consideration the correlation that exists between $\mathbf{x}$ and $\mathbf{z}$.

Because the two vectors between which the distance is measured may be of unequal length, a linear time normalization is essential. One disadvantage of linear time warping is that it does not compensate for the varying length of utterances within a word boundary. Usually stop consonants are shorter while dipthongs and glides may have longer durations. This problem may be solved by nonlinear dynamic time warping (DTW) so that the variability in duration is compensated in a dynamic manner, usually by dynamic programming [75], [76].

Finding the minimum metric distance from a new vector to a stored prototype is equivalent to finding the maximum dot product correlation between the new vector and the prototype. Finding the best prototypes for this measurement then can be viewed as determining weights for a linear combination of the input features that will be maximum for the correct class, which may be often determined by a discriminant analysis. Linear discriminant functions generate linear decision surfaces. In two dimensions, the surface is a line and in three dimensions, it is a plane. Nonlinear decision surfaces, such as quadratic surfaces, may be obtained by utilizing an artificial neural networks (ANN), which is a class of classifiers used in speech recognition motivated by the biological nervous system. It is a nonprogrammed (nonalgorithmic) adaptive information processing system that develops associations between objects in response to their environment. That is, they are capable of learning from a large set of examples [77].

2.7.2 Acoustic modelling using HMM

A more robust technique than the deterministic approach for speech recognition is the statistical method using a Hidden Markov model (HMM) [78], [68], [2], [79] which to-date has become the state-of-the art technique for ASR. Since speech has numerous variabilities which are usually unpredictable, it is possible to model speech as a context dependent random process and then use statistical tools to analyze it. Speech models used for this purpose are the hidden Markov models (HMM) which allow parameter estimation of hidden or unobservable speech feature parameters from known and observable parameters. The Markov chain rule is one of the most widely used models describing class dependence in a classification task. For a sequence of classes $\omega_{i_1}, \omega_{i_2}, \ldots$, the first-order Markov model assumes that

$$P(\omega_{i_k}|\omega_{i_{k-1}}, \omega_{i_{k-2}}, \ldots \omega_{i_1}) = P(\omega_{i_k}|(\omega_{i_{k-1}}). \tag{2.21}$$

In other words, given that the observations $\mathbf{o}_{k-1}, \mathbf{o}_{k-1}, \ldots, \mathbf{o}_1$ belongs to classes $\omega_{i_{k-1}}$,
$\omega_{i_{k-2}}, \ldots, \omega_{i_1}$, respectively, the probability of the observation $\mathbf{o}_k$, at

stage k, belonging to class ω_{i_k} depends only on the class from which observations $\mathbf{o}_{k-1}$, at stage $k-1$, has occurred. Having observed the sequence of the feature vectors $O{:}\mathbf{o}_1,\mathbf{o}_2,\ldots,\mathbf{o}_N$, it is sufficient to classify them in the respective sequence of classes $\Omega_i{:}\omega_{i_1},\omega_{i_2},\ldots$, so that the quantity

$$p(O|\Omega_i)P(\Omega_i) = P(\omega_i)p(\mathbf{o}_1|\omega_{i_1}) \prod_{k=2}^{N} P(\omega_{i_k}|\omega_{i_{k-1}})p(\mathbf{o}_k|\omega_{i_k}) \qquad (2.22)$$

becomes maximum.

To classify an observed vector sequence O to a particular class sequence Ω, transitions are made from one class ω_i to another class ω_j with a fixed and known probability $P(\omega_j|\omega_i)$ as observation vectors are obtained in sequence. These probabilities depend only on the respective class transitions from state i to state j and not on the stage at which they occur. It is also assumed that conditional probability density $p(\mathbf{o}|\omega_i)$, $i=1,2,\ldots,M$ are known to the model. The path of successive transition (class) transitions that maximizes Eqn. (2.22) is the optimal one in which the observation vector will be classified. The states of the Markov chain may be observable. However, the sequence in which the different states are visited by successive observations is itself the result of another stochastic process, which is hidden to us, and the associated parameters describing it can only be inferred by the set of the received observations. These types of Markov models are known as hidden Markov models (HMMs) [78].

2.7.3 HMM for word recognition

For applications in speech recognition, in contrast to template matching approach of pattern classifier stated in Sec. 2.7.1, the HMM uses statistical modelling, a vocabulary database and grammar rules to select the highest probability outcome from a sequence of observation feature vectors. When the vocabulary is small, the entire word can be modeled as a single unit. But such an approach is not practical for large vocabularies. For such cases, word models must be built up from subword units such as phonemes. Effects of linguistic context at the acoustic phonetic level are usually handled by training separate models for phonemes in different contexts. This is called context dependent acoustic modelling [80], [2]. Word level variability is usually handled by allowing alternate pronunciations of words in a pronunciation network.

In HMM-based word recognition [68], it is assumed that the sequence of observed speech vectors $O = o_1, \ldots, o_T$ corresponding to each word is generated by a Markov model. Fig. 2.10 shows a simple left-to-right HMM with five states and three emitting states. An observation $p_j(.)$ vector is emitted each time there is a change or transition of state j, which is controlled by the transition probabilities a_{ij} associated with each state j. A state corresponds to an acoustically stable region or a number of continuous number of frames with stationary observation vectors. In continuous density models, each output observation probability distribution $p_j(.)$ is represented by a mixture Gaussian density [1] given as

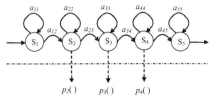

$$p_2(\,) \qquad p_3(\,) \qquad p_4(\,)$$

FIGURE 2.10: A simple left-to-right HMM with 5 states S_j and 3 emitting states with probabilities p_j $(0 < j \leq 5)$ and having a Gaussian mixture distribution. An observation vector $p_j(.)$ is emitted whenever there is a change or transition of state j, depending on the transition probabilities a_{ij}.

$$p_j(o_t) = \prod_{s=1}^{S} \left[\sum_{m=1}^{M_s} c_{jsm} \aleph(o_{st}; \boldsymbol{\mu}_{jsm}, \boldsymbol{\Sigma}_{jsm}) \right]^{\gamma_s} \qquad (2.23)$$

where M_s is the number of mixture components in stream s, c_{jsm} is
the weight of the m-th component and $\aleph(.; \boldsymbol{\mu}, \boldsymbol{\Sigma})$ is a multivariate
Gaussian with mean vector $\boldsymbol{\mu}$ and covariance matrix $\boldsymbol{\Sigma}$, given in its
general form by

$$\aleph(o; \boldsymbol{\mu}, \boldsymbol{\Sigma}) = \frac{1}{\sqrt{(2\pi)^n |\boldsymbol{\Sigma}|}} \exp^{-\frac{1}{2}(o-\boldsymbol{\mu})'\boldsymbol{\Sigma}^{-1}(o-\boldsymbol{\mu})} \qquad (2.24)$$

where n is the dimensionality of O. The exponent γ_s is a stream
weight also referred to as the codebook exponent [68].

If the variable O's dimensionality, n, is large (eg 40, for speech
recognition problems), then the use of full (non diagonal) covariance
matrices $\boldsymbol{\Sigma}$ would involve a large number of parameters (on the order of $(M \times D^2)$). To reduce such a number, a diagonal covariance
matrix is used, provided the data is made uncorrelated through some
decorrelation process, such as DCT [69] or cepstral transformation [5].

HMM recognition is based on the concept of the Bayesian maximum
a posteriori (MAP) classifier, which expresses the posterior probabilities in terms of the a priori probabilities and the likelihood. The
likelihood can be estimated iteratively from a large set of training
data. The Bayesian classifier is given as

$$P(w_k|O) = \frac{P(O|w_k)P(w_k)}{P(O)} \qquad (2.25)$$

which states that for a given set of prior probabilities $P(w_k)$ and $P(O)$,
the most probable spoken word, $P(w_k|O)$, depends on only the likeli-
hood $P(O|w_k)$. In Eqn. (2.35), $P(w_k)$ is the probability of class w_k,
also called a priori probability of class k, since it can be evaluated before O has been observed, and $P(O)$ is the probability of observation

vector, which is same for all classes.

A feature vector O can be optimally classified into a class k by using a maximum a posteriori (MAP) decision rule which assign O to class k if

$$P(w_k|O) > P(w_j|O), \qquad j = 1, \ldots, k, \quad \text{for} \quad \forall \; j \neq k. \qquad (2.26)$$

In other words, the class with the maximum posterior probability is chosen. Besides posterior probabilities, $P(w_k|O)$, often it is easier to generate estimates of likelihood $P(O|w_k)$. In Bayes's rule, in fact, it is not required to compute $P(O)$, since $P(O)$ is a constant for all classes. A likelihood ratio of posterior probabilities is defined [2] as

$$\frac{P(w_k|O)}{P(w_j|O)} = \frac{P(O|w_k)P(w_k)}{P(O|w_j)P(w_j)} \qquad (2.27)$$

According to Bayes's decision rule, w_k is selected over w_j if the ratio is greater than 1. This case is equivalent to assigning the observation O to class k if

$$\frac{P(O|w_k)}{P(O|w_j)} > \frac{P(w_j)}{P(w_k)} \qquad \text{for} \quad \forall \; j \neq k \qquad (2.28)$$

As for the re-estimation case, the direct computation of likelihoods leads to underflow, hence, log likelihoods are used instead. Taking the log of the likelihood, the rule states that w_k is chosen over w_j if

$$\log P(O|w_k) + \log P(w_k) > \log P(O|w_j) + \log P(w_j). \qquad (2.29)$$

The MAP classification based on likelihoods is then equivalent to choosing a class to maximize a statistical discriminate function

$$\operatorname*{argmax}_{k} \; [\log P(O|w_k) + \log P(w_k)]. \qquad (2.30)$$

The likelihood densities are unknown and must be estimated from a large set of training data containing examples of each class w_k. Assuming parametric form of the distribution of a class, estimators of $P(O|w_k)$ are trained for each class which results in the learning of some set of parameters, Θ, for the distribution, which is used during classification to estimate the required probabilities. Maximum likelihood (ML) estimation procedures are used to learn the parameters that will give the largest values for these quantities. For a density represented as a weighted sum or mixture of Gaussian densities, with means, covariances, and mixture weights to be learned from the data, the parameters are usually estimated with an iterative solutions (Baum-Welch re-estimation) using the expectation maximization (EM) algorithm. In the EM algorithm, the parameter estimation is done by incorporating variables that are not observable (that is, hidden) but are assumed to be part of the model that generated the data. The densities are estimated by taking an expectation of the logarithm of the joint density between the known and unknown components, and then this function is maximized by updating the parameters that are used in the probability estimation. This process is then reiterated as required. It can be shown that if this expectation is maximized, so too is the data likelihood itself.

For N observations and M mixture components (hidden) from which each data sample came, the expected value for the log joint density for the observed and the hidden variables, $\mathbf{Q}$, which is to be optimized, is given in [2] as

$$Q = \sum_{m=1}^{M} \sum_{n=1}^{N} P(m|o_n, \Theta_{\text{old}}) \log P(m|\Theta) + \sum_{m=1}^{M} \sum_{n=1}^{N} P(m|o_n, \Theta_{\text{old}}) \log P(o_n|m, \Theta)$$

$$(2.31)$$

where Θ is the variable to be optimized and Θ_{old} is the parameter used to generate the distribution with which the expectation is iterated.

A possible method to train the parameters for this density is to assume that each mixture component $P(o|m)$ is a Gaussian with mean μ_m and variance σ_m^2. The optimum value of the means and the variances, computed iteratively by the EM algorithm, is given as

$$\mu_m = \frac{\sum_{n=1}^{N} P(m|o_n, \Theta_{\text{old}}) o_n}{\sum_{n=1}^{N} P(m|o_n, \Theta_{\text{old}})} \qquad (2.32)$$

$$\sigma_m^2 = \frac{\sum_{n=1}^{N} P(m|o_n, \Theta_{\text{old}})(o_n - \mu_m)^2}{\sum_{n=1}^{N} P(m|o_n, \Theta_{\text{old}})}. \qquad (2.33)$$

2.7.4 Recognition using Viterbi decoding

In the word recognition process, an optimal string search match between the transcription output (label sequence) of the recognized word and a original reference transcription is performed. When dynamic programming is used to get the best match between the data and the statistical model, the resulting best path calculation is called Viterbi decoding [68]. The algorithm is shown in in Fig. 2.11, which can be visualized as finding the best path through a matrix where the vertical dimension represents the states of the HMM and the horizontal dimension represents the frames of speech (time). The black solid dots at matrix intersections represents the log probability of observing that frame at that time and each path between the two dots corresponds to a log transition probability. The log probability of any path for a model is computed simply by summing the log transition probabilities and the log output probabilities along the path. The observed utterance is assigned to the model which has the highest probable path. Usually it does not take into account any boundary timing information. This concept of path in Viterbi decoding has the advantage that it can be generalized very easily to the case of continuous speech recognition.

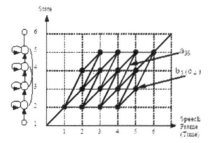

FIGURE 2.11: Viterbi algorithm to find the best match for isolated word recognition [68].

The optimal string match is the label alignment for the labels in the reference transcriptions which has the best match, usually expressed as the percent correct of the recognition process, given as

$$\text{Percent Correct} = \frac{N - D - S}{N} \times 100 \qquad (2.34)$$

where N is the total number of labels in the reference transcriptions, and S, and D are the total number of substitutions and deletions errors, respectively.

A more accurate representation is the percent accuracy which takes into account the insertion errors,

$$\text{Percent Accuracy} = \frac{N - D - S - I}{N} \times 100 \qquad (2.35)$$

where I is the total number of insertions errors.

2.8 Auditory models for speech recognition

Of the human auditory system, consisting of the peripheral and the central auditory system, the function of the peripheral auditory system is better understood. Therefore, conventional auditory models primarily mimic the functionalities and psychoacoustics of the inner ear. It simulates the transformation of the mechanical vibrations of the basilar membrane into neural representations through a series of nonlinear transformations such as compressive nonlinearity, synaptic adaptation, two-tone suppression and masking effects. Primary applications of auditory models include speech and hearing research, speech processing for hearing aids and cochlear implants. However, auditory models have been applied in automatic speech recognition in the past years, primarily motivated by the exceptional ability of human speech perception under adverse conditions [16], [19], [21], [24], [7]. A computational auditory model simulates the human auditory system and its perceptual characteristics from which reliable speech features can be extracted for suitable processing in a recognition system.

However, most of the standard speech parameterizations such as the MFCC and the PLP utilize the psychoacoustics and physiology of human auditory system in one form or another. For example, the transformation of speech vectors to spectral magnitudes emphasizes the role of the BM as a spectral analyzer [39]. The filterbank decomposition techniques and the use of mel or Bark scales (critical bands), simulates the tonotopic processing in the BM [41]. Static compression with log or cube-root functions, and spectral smoothing are consistent with the perceptual features of auditory compression [70]. The spectrum is pre-emphasized to implement the functionalities of the outer and the middle ear, that is, the unequal sensitivity of human hearing at different frequencies corresponding to the equal loudness curves of perception [9]. In the filterbank analysis, the progressively higher bandwidths at higher frequencies give poorer frequency resolution which simulates the loss of synchrony and the ability to phase lock at higher frequencies. RASTA processing has some relation to the models of forward temporal masking [71]. However, these systems when deployed in real applications, still lack robustness in performance. Robustness in speech recognition refers to maintaining good recognition accuracy even when the quality of the input speech is degraded, or when the acoustical, articulatory, or phonetic characteristics of speech in the training and testing environments differ [80]. As speech recognition and spoken language technologies are being transferred to real applications in diverse and hostile environments, the need for greater robustness in recognition technology is becoming increasingly important.

Fig. 2.12 shows connected digit utterances '139o' (upper panel) mixed with four different types of noise obtained from the NOISEX 92: Gaussian stationary (white noise), non-Gaussian and non-stationary (factory noise), non-Gaussian and pseudo-stationary (babble noise) and non-Gaussian stationary (Volvo noise). In each case, the speech is corrupted with the respective noise at 10 dB SNR. The lower panel shows the corresponding power spectrum of the clean speech and the power spectrum of the noisy speech. The figures show the dependence of frequency response of speech corrupted by noise on the noise types, particularly the high and low frequency discrimination in presence of the stationary Gaussian noise and the non-stationary real-world noise.

Several speech enhancement techniques based on noise subtraction have been developed to reduce the effects of noise. Some researchers have tried to model the noise itself or train the model parameter adapted in accordance with the noisy conditions [81]. Additionally, statistical estimation techniques using Wiener and Kalman

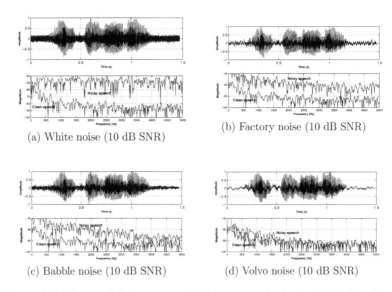

(a) White noise (10 dB SNR)

(b) Factory noise (10 dB SNR)

(c) Babble noise (10 dB SNR)

(d) Volvo noise (10 dB SNR)

FIGURE 2.12: Connected digit utterance '139o' (upper panel) mixed with (a) white, (b) factory, (c) babble and (d) Volvo noise (10 dB SNR). The corresponding power spectra are shown in the lower panels for clean speech and the noisy speech.

filtering have been used to estimate optimum signals buried in noise. Speaker dependence has been addressed by including speaker adaptation into the training phase using maximum likelihood linear regression (MLLR) [82].

In recent times, there has been a resurgence in the tendency of the researchers to include specific perceptual-based features in speech recognition. Some of the research works done in the field of auditory processing for speech recognition are summarized in Table 2.3. Several observations are made from the table. Although Zwicker [41] proposed the critical bands in 1957, it is noted that the first attempt of applying the filterbank approach to speech recognition, was in the mid 70's. Serious efforts for speech recognition using the filterbank approach and the acoustic properties of speech sound (acoustic preprocessing) related to perception were first attempted in the late 70's.

These recognition experiments were based on classical classification
techniques such as the Euclidean distance spectrum comparison, since
use of standard HMM techniques in speech recognition was not available at that time. The experiments were mostly performed in clean
condition since these classification techniques were not particularly
suited to noisy conditions. In subsequent experimentations, perceptual features have been employed in a variety of speech processing
applications [9], [11], [16], [19]. Some of these models have been applied to ASR, and front-ends based on the auditory system have been
shown, in specific cases, to perform better than the more conventional
signal processing schemes for speech recognition tasks [16]. However,
the use of perceptual features for speech recognition has not resulted
in improved results in many cases. For example, Bloomberg *et al.* [20]
reported a series of experiments in which bark, phon, sone, and other
perceptually motivated representations of speech spectra were used
in an isolated-word recognition system. The results were disappoint-
ing, as each additional level of sophistication in the model produced
larger recognition-error rates. The poor performance with auditory
features was the consequence of both suboptimal choice of the features themselves and the lack of a good match between their characteristics and the characteristics of the speech recognition to which
they are input [122]. The state of the art HMM technique for perceptual speech recognition was first widely introduced in the early
1990's [24], [22], [27]. In general, such application have proved more
robust than conventional speech recognition.

The key point raised in the literature review is the transition of
speech recognition from speech processing using known properties of
human auditory perceptions. It has been shown that spectral processing of speech may lead to improved performance with specific perceptual features, such as synaptic adaptation [27]. Other perceptual

features such as two-tone suppression has not been investigated for speech recognition. Moreover, there are a very few research carried out using time-domain filtering and temporal processing of speech. Such investigations are conducted in this research and stated in the subsequent chapters of this book.

Based on the research to-date as stated in Table 2.3, three methodologies are employed in this research as stated in Sec 1.1. Firstly, we have emphasized on temporal processing of speech which is more consistent with human perceptual behaviours, rather than spectral domain processing, which emphasizes more on the frequency content. Many of the experimental investigations utilize temporal processing of speech using time domain filtering and a zero-crossing algorithm for feature extraction. Secondly, the performances with the psychoacoustic effects were evaluated utilizing the state of the art continuous density HMM using clean and noisy speech, and measured in terms of the word recognition rates. Thirdly, for the feature extraction process, larger time windows and filters with appropriate time constants consistent with human perception are utilized. Based on these methodologies, a HMM standard baseline performance for speech recognition with perceptual features was also defined for establishing a basis of comparison, which is shown in Table 2.4.

It is observed from the literature review that accumulated knowledge of the perceptual properties of the inner ear cells have led to the evolution of fairly precise models [86] for speech processing. These models are not suitable for ASR primarily due to computational complexities. Even if it is possible to extract features for ASR, the detailed implementations are not relevant for ASR applications. Therefore, simplified models have been developed in which the basic functions of the perceptual features are incorporated to study their effects on ASR performance. Starting from the simplified model, the model may be further improved with careful choice of parameters, so that a better model for ASR applications may be developed. The work herein is basically application and extension of this approach. The simplified models significantly improves performance [27], which is sufficient to justify this approach. The research also includes several additional perceptual features applied to speech recognition, including development of several novel processing algorithms.

A classical frame based method embedded in a hidden Markov model (HMM) technique is used to evaluate temporal features/effects represented by the ZCPA model outputs. The classical frame-based framework (typically 25 ms) is based on the assumption that the features are stationary (or quasi stationary) within the frame window (duration). There may be some model-feature mismatch between the frame-based approach and auditory events like rapid synaptic adaptation, and also due to the wider window lengths (30 ms to 120 ms) employed in auditory processing, with some loss of assumption of stationarity. As a result, a frame-based approach for HMMs in auditory processing has always been a compromise. Nevertheless, frame-based HMM technique for ASR in auditory models have been shown to provide improved performances as demonstrated by several researchers [16], [24], [7], [27].

TABLE 2.3: A review of research to-date on application of perceptual features for speech recognition.

Reference /year	Methodology-perceptual processing	Recognition process	Results	Comments
E. Zwicker et al. [17] /1979	Critical band filterbank, loudness,pitch,roughness, subjective duration,spectral momenta related to loudness pattern.	Detection/classification of syllable peaks,vowel recognition using discriminant functions	23 % error rate on isolated word classification.	Classification of consonant clusters identified as main source of error, not tested in noise
C. L. Searle et al. [18] /1979	1/3 octave filterbank envelope detectors, logarithmic amplifier.	Discrimination of stops, phoneme recognition using discriminant analysis program.	77 % of stops correctly detected.	VOT detected as the most significant feature, not tested with added noise.
M. Bloomberg et al. [20] /1984	Critical band filterbank, features extracted from , loudness, (phons and sones).	Isolated word recognition.	Disappointing performances.	-
S. Seneff [83] /1984	Critical band filterbank, compression/adaptation, HWR,generalized synchrony detector (GSD), for detecting periodicities.	Applied to pitch estimation of isolated words	NR	-
M. Hunt et al. [15] /1986	Cascade of second order sections,compression/ adaptation/AGC, modified GSD for freq., masking replication.	Isolated digit (male), dynamic programming for word comparison (Euclidean distance) spectrum comparison.	Performed better than MFCC in clean, noise and linear distortions.	Sensitivity to interfering steady periodic signals identified as weak point.
R. Lyon et al. [11] /1988	Cascade of second order sections digital filters, adaptation, AGC.	NR	NR	Implemented for speech processing with objective for speech recognition
S. Seneff [9] /1988	Critical band filterbank, HWR, short-term adaptation, rapid AGC, GSD and mean rate output.	NR	NR	-do-
J. R. Cohen et al. [16] /1989	Critical band filterbank , compressive power law transformation, reser--voir type adaptation.	100 sentence training data, 50 sentence test data, ML decision rule using acoustic (label) and language model.	Error rates and decoding times substantially lower than the filterbank model (3.9% vs. 6.3%).	Not tested with added noise other than picked up by microphone.
O. Ghitza [19] /1994	Multibandpass nonlinear filter, ensemble interval histogram (EIH) (multilevel).	Isolated word speaker dependent dynamic rhyme test (DRT), maximum likelihood decision rule with HMM word models.	EIH was more robust to noise than Fourier power spectrum method.	Errors made main in the presence of voicing and absence of sustention.
C.Jankowski et al. [21] /1995	GSD and EIH auditory models compared with mel filterbank based cepstral front end.	Tested on TI-105 isolated word database, HMM recognizer in clean and babble noise.	The auditory model provided error rates as much as 4 % lower than a mel filterbank.	The study also showed that mel filterbank cepstra outperformed LPC spectra.
S. Sandhu et al. [23] /1995	The EIH and the conventional mel cepstral analysis was compared for phone classification.	HMM recognizer, TIMIT database for clean telephone and reverberated speech with static and dynamic features.	EIH was more robust to noise than mel cepstral representation.	Contribution of dynamic features EIH was much smaller compared to the mel filterbank.
B. Strope et al. [24] /1997	Dynamic model with a logarithmic adaptation stage (AGC) based on forward masking data.	Isolated digits DTW (Itakura path constraint) and HMM recognizer with forced Viterbi alignment.	Improvements in robustness to background noise compared to the MFCC, LPC and RASTA based front-ends.	-
D. S. Kim et al. [7] /1999	Auditory model for speech recognition, FIR filterbank,single-level EIH, peak detector.	Isolated 50 Korean words using discrete density HMM and 256 word codebooks.	Improved robustness in various types of noise compared to LPCC, MFCC, PLP, EIHC.	ZCPA is more robust in white noise than in real-world noise types.

Reference /year	Methodology-perceptual processing	Recognition process	Results	Comments
Tchorz et al. [22] /1999	Gammatone filterbank, half-wave rectification, amplitude compression by casacde offive non linear adaptation loops.	Speaker independent, isolated word recognition using a HMM with additive noise.	Adaptation system more robust in noise than mel scale cepstral eatures.	It was identified that the adaptive compression stage was the most important processing stage.
L. Bu et al. [84] /2000	Masking effects, MAF-curve normalization, mel-scale resampling, phonetic feature extraction by similarity measures.	Mandarin vowel, phoneme recognition with continuous density HMM.	Phonetic features increased recognition robustness to speaker variability and added noise.	-
M.Holmberg et al. [27] /2006	Simplified synaptic short-term adaptation introduced in MFCC by high-pass first order IIR flter.	Aurora 2 and 3 (added noise) with continuous density HMM recognition.	Improved ASR performance compared to baseline MFCC, MFCC with RASTA, CMS, Weiner filtering.	A filter time constant of 240 ms for adaptation was found appropriate for ASR.
C. Kim [122] /2006	Zhang-Carney AN model implemented with auditory synchrony and mean rate processing.	CMU Sphinx III with DARPA RM and WSJ database in broadband and transient noise.	Synchrony with mean rate more robust than MFCC or mean rate only.	AN model identified as more robust than a model with Gammatone filters, compressive nonlinearity.

NR-Not reported

TABLE 2.4: Comparison of research in Table 2.3 based on a HMM standard baseline performance for speech recognition with perceptual features.

Reference /year	Short-term processing	Perceptual filterbank	Compre-, ssion AGC	Adaptation forward/si-multaneous masking	Two-tone supp-ression	Temporal processing	Decorre-lation	HMM recognition
Zwicker et al. /1979	NR	Yes	NR	NR	NR	Yes	NR	NR
Searle et al. /1979	Yes	Yes	Yes	NR	NR	Yes	NR	NR
Bloomberg et al. /1984	NA	NA	NR	NR	NR	Yes	NA	NR
Seneff /1984	NR	Yes	Yes	Yes	NR	Yes	NR	NR
Hunt et al. /1986	NR	NR	Yes	Yes	Yes	Yes	NR	NR
Lyon et al. /1988	NR	NR	Yes	Yes	Yes	Yes	NR	NR
Seneff /1988	NR	Yes	Yes	Yes	NR	Yes	NR	NR
Cohen /1989	NR	Yes	Yes	Yes	NR	Yes	NR	NR
Ghitza /1994	Yes	Yes	Yes	NR	NR	Yes	Yes	Yes
Jankowski /1995	NR	Yes	NR	Yes	NR	Yes	Yes	Yes
Sandhu et al. /1997	Yes	Yes	NR	NR	NR	Yes	Yes	Yes
Strope et al. /1997	Yes	Yes	Yes	Yes	NR	Yes	Yes	Yes
D.S. Kim et al. /1999	Yes	Yes	Yes	NR	NR	Yes	Yes	Yes
Tchorz et al. /1999	Yes	NR	Yes	Yes	NR	Yes	NR	Yes
L. Bu et al. /2000	Yes	Yes	NR	Yes	NR	NR	Yes	Yes
Holmberg et al. /2006	Yes	Yes	Yes	Yes	NR	NR	Yes	Yes
C. Kim et al. /2006	NR	Yes	Yes	NR	NR	NR	NR	NR

NR-Not reported   NA-Not available

Chapter 3

Synaptic adaptation in a temporal auditory model

In this chapter, the role of the perceptual property of synaptic adaptation in speech recognition is investigated. The auditory system is sensitive to abrupt stimulus changes and the dynamic and transient component in speech are particularly critical to speech perception [24], [26]. Auditory adaptation, as introduced in Sec. 2.3.4, is a dynamic mechanism of the human auditory system. This is characterized by an increased neural discharge rate at signal onset, with a rapid decrease within the first 3 ms giving rise to rapid synaptic adaptation, and a gradual decay to a steady-state value within about 50 ms after the onset transient, producing the short-term synaptic adaptation [54], [31]. Adaptation in the AN results primarily from the release of transmitter substance from the hair cells into the synaptic cleft due to signal onset and the subsequent transmitter depletion of Ca^{2+} in the synaptic cleft [85], [86]. After the release of neurotransmitters by the hair cells, the voltage-sensitive Ca^{2+} channels located close to the synapses at the basal part of these cells open upon depolarization of the cell membrane. This process empties the "readily releasable pool" (RRP). The cleft is filled up at a rate that is proportional to the concentration gradient across the membrane. Some transmitter within the cleft is lost which gives rise to the slow depletion of the neurotransmitters. Some flow back through the cell membrane for recycling (reuptake) and are reprocessed to fill up the RRP. At the beginning of an acoustic stimulus, plenty of vesicles are available to fuse, causing a strong initial auditory nerve response. As the RRP is refilled at a lower rate than the initial vesicle fusion rate due to the loss in the synaptic cleft, it depletes with time. Auditory nerve activity is, thus, depressed shortly after the stimulus onset transient and during sustained stimuli, producing the synaptic adaptation effect. This process is shown in Fig. 3.1.

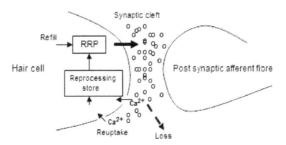

FIGURE 3.1: Schematic representation of the synaptic adaptation process in the synapses between the inner hair cells and an AN using the vesicle pool dynamics showing the reprocessing and the loss of the neurotransmitter substances in the synaptic cleft [27].

Adaptation accentuates signal onsets by following a high initial discharge rate. A rapid synaptic adaptation component enhances voicing, and short-term synaptic adaptation is found to improve the immunity of the system to stationary noise [87]. This principle is used in RASTA processing of speech and has been shown to improve the robustness of the system [71]. However, the main difference is that RASTA processing is a bandpass modulation filtering operating on the logarithmic spectrum, which completely suppresses dc modulations, whereas the auditory nerve shows a sustained discharge rate to steady stimuli.

The perceptual features of synaptic adaptation have been employed in ASR by several researchers. Cohen [16] implemented an ASR front-end utilizing a Bark-scaled filterbank in which a compressive power law transformation was applied to the output of each filter approximating loudness scaling. A reservoir-type synaptic adaptation was used to model the neural discharge rate. The model was tested using a 100 sentence training corpus and a 50 sentence test corpus with a 5000 word vocabulary. It was shown that the error rates and decoding times were substantially lower with the auditory model than with a mel filterbank implementation (3.9% vs. 6.3%).

Seneff's generalized synchrony detector (GSD) model [9] incorporated an IHC-synaptic processing which consisted of a saturating half-wave rectification, short term synaptic adaptation, synchrony suppression at higher frequencies and rapid adaptation as observed from psychophysical measurements. Jankowski [21] designed a recognizer using Seneff's GSD auditory model and compared the word error rates with a conventional mel filterbank. Recognition tests were performed on clean and noisy speech using the the TI-105 isolated word corpus and a HMM recognizer. The auditory model provided error rates as much as four percent lower than the mel filterbank. However, Perdigao *et al.* [88] found the Seneff's GSD model to be susceptible to noise, in contrast to the finding in [21], but observed that a simplified model of synaptic adaptation generally improved recognition performances.

Strope *et al.* [24] described a dynamic model with a logarithmic adaptation stage based on forward masking data. The synaptic adaptation mechanism implemented was a modified form of automatic gain control (AGC), which added an exponentially adapting linear offset to the logarithmic intensity. It showed improvements in robustness to background noise when used as a front-end for DTW and HMM based recognition. Tchorz *et al.* [22] designed a front-end for automatic speech recognition utilizing both spectral and temporal properties of sound processing in the auditory system. This included a gammatone filterbank, envelope extraction by half-wave rectification, and amplitude compression by an adaptation circuit which consisted of five consecutive nonlinear adaptation loops, with each loop having an increasing time constant. The model was evaluated in speaker independent, isolated word recognition using a HMM recognizer in different types of additive noise. It was found to be more robust in noise than mel scale cepstral features. It was also determined that the adaptive compression stage which enhanced temporal changes of the input signal appeared to be the most important processing stage toward robust speech representation in noise.

Holmberg *et al.* [27] incorporated a simplified model of synaptic adaptation for ASR into the MFCC feature parametrization as a competitive strategy to RASTA and the cepstrum mean subtraction (CMS), and as an incremental addition to conventional MFCC feature extraction. The purpose of the model used is to motivate a simplified synaptic adaptation stage that in fact may describe aspects of synaptic adaptation, a ubiquitous property of auditory synapses, not only in the auditory nerve, but also of processing stages at higher levels along the auditory pathway. A first-order infinite impulse response (IIR) high-pass filter was used to represent the decaying exponential effects of the short term synaptic adaptation. The synaptic adaptation filtering was inserted into MFCC feature extraction just after the calculation of logarithmic mel spectra by summing (with equal weights) a temporally highpass filtered version of the logarithmic mel spectra with the original logarithmic mel spectra. Evaluated with AURORA 2 and 3 corpus (connected TIDIGITS spoken by both male and female speakers)and a HMM recognizer, it was determined that a short-term synaptic adaptation time constant of 240 ms is appropriate which is consistent with estimates for synaptic adaptation in the human auditory nerve as determined by Spoor *et al.* [89]. With this time constant, the recognition experiments showed improved ASR performance compared to the baseline MFCC, and MFCC in combination with RASTA and CMS processing, and both in comparison to and in combination with Wiener filtering. For example, using the AURORA2 recognition task, the word error rate is reduced by 46% relative for clean training, and by 17% relative for multicondition training. With Wiener filtering, the improvements are higher, that is 51.6% and 18.9%, respectively.

Our approach is similar to the one used by Holmberg *et al.* [27], but differs in three aspects. Firstly, it operates in the time domain rather than in the spectral domain. The advantages of time domain over spectral domain in auditory processing are stated in Sec. 1.1. Specifically, rapid synaptic adaptation, which enhances the temporal fine structure of speech signals, cannot be implemented in the spectral domain because this fine temporal structure is removed by the spectral feature extraction process [27].

Secondly, our experimental setup for temporal processing utilizes a zero-crossing algorithm for speech feature extraction from the periodicities of the input signal. For the implementation phase, we selected the zero-crossings with peak amplitudes (ZCPA) auditory model [7] with simple FIR filters as cochlear filters, with only 16 filters in the filterbank for computational efficiency. The base ZCPA temporal auditory model do not provide any synaptic processing through which dynamic or transient properties of speech are emphasized as observed in the auditory system. The proposed method extends the capabilities of the ZCPA by implementing synaptic temporal synaptic adaptation in it.

Thirdly, forward masking can be viewed as a consequence of auditory adaptation. In particular, recovery from synaptic adaptation may be responsible for temporal (forward) masking observed in psychoa-coustic experiments [24], [55]. Ghulam *et al.* [90] combined forward and backward masking with the pitch synchronous ZCPA, which was half-wave rectified and centre clipped. The addition of masking effects was found to improve recognition rates. Our approach differs from this method in that we do not utilize a half-wave rectification, because it introduces substantial higher order harmonics of the formant frequen-cies [9], which degrades the formant detection, and hence the ASR performance. Moreover, half-wave rectification removes the negative portion of the signal and hence, histogram construction utilizing zero crossings is not possible. The mean is also raised from the zero value, but level values higher than the zero level result in higher sensitivity in the estimated intervals and frequencies for the ZCPA [7], degrading the ASR performance.

3.1 Temporal processing of speech with zero-crossing algorithm

For a real-valued zero mean stationary Gaussian random process $Z_t, t \in T$ for all $t_1, \ldots, t_N \in T$, the number of zero-crossings D, can be defined in terms of a clipped process X_t for discrete time t as

$$D = \sum_{t=1}^{N} [X_t - X_{t-1}]^2, \qquad 0 \leq D \leq N - 1 \qquad (3.1)$$

where

$$X_t = \begin{cases} 1 & \text{if} \quad Z_t \geq 0 \\ 0 & \text{if} \quad Z_t < 0 \end{cases}$$

for $t=1,\ldots,N$. If we define $\boldsymbol{\mu}=E\{Z_t\}$ as the mean vector ($E\{.\}$ denotes the expected value), and γ_k as the covariance matrix as a function of the lag k such that

$$\gamma_k = \text{Cov}(Z_t(k)) = E\{(Z_t - \boldsymbol{\mu})(Z_{t+k} - \boldsymbol{\mu})\} \tag{3.2}$$

then normalizing the covariance γ_k by γ_0 so that it is invariant to scaling, we may express

$$\rho_k = \frac{\gamma_k}{\gamma_0} \qquad k = 0, \pm 1, \ldots. \tag{3.3}$$

which is the autocorrelation function such that $|\rho| \leq 1$. Using the Wiener-Khintchin theorem, it is shown in [91] and [29] that a monotone increasing function, $F(\omega)$ exists (ω being the angular frequency with a maximum value of π), such that

$$\gamma_k = \int_{-\pi}^{\pi} \cos(k\omega) dF(\omega) \tag{3.4}$$

where F is the spectral distribution function (cumulative) of the random process Z_t. For $k=1$,

$$\rho_1 = \frac{\gamma_1}{\gamma_0} = \frac{\int_{-\pi}^{\pi} \cos(\omega) dF(\omega)}{\int_{-\pi}^{\pi} dF(\omega)}. \tag{3.5}$$

Since Z_t is Gaussian with zero mean with $E\{X_t\} = \frac{1}{2}$, an expression for $E\{D\}$, the expected value of the zero-crossings of Z_t may be determined. It is shown in [29] that

$$E\{X_t X_{t-1}\} = \frac{1}{4} + \frac{1}{2\pi} \sin^{-1}(\rho_1). \tag{3.6}$$

The number of zero-crossings D is the same as the number of symbol changes in the X series and is obtained by summing all the $(X_t - X_{t-1})^2$ values. Hence, the expected value of the zero-crossings

corresponding to the symbol changes in the series X may be expressed
as

$$E\{D\} = E\left\{\sum_{t=1}^{N}[X_t - X_{t-1}]^2\right\} = \sum_{t=1}^{N}\left[E\{X_t\}^2 - 2E\{X_t X_{t-1}\} + E\{X_{t-1}\}^2]\right]$$

which may be expressed as

$$E\{D\} = (N - 1)\left(\frac{1}{2} - \frac{1}{\pi}\sin^{-1}(\rho_1)\right). \tag{3.7}$$

By rearranging terms,

$$\sin^{-1}(\rho_1) = \frac{\pi}{2} - \frac{\pi E\{D\}}{N - 1}$$

$$\text{or} \quad \rho_1 = \cos\left(\frac{\pi E\{D\}}{N - 1}\right). \tag{3.8}$$

Equating Eqn. (3.5) and Eqn. (3.8), and since $F(\omega)$ is continuous at the origin and
symmetrical, we may express for only the positive frequencies

$$\cos\left(\frac{\pi E\{D\}}{(N-1)}\right) = \frac{\int\limits_0^\pi \cos(\omega)dF(\omega)}{\int\limits_0^\pi dF(\omega)} \tag{3.9}$$

where $F(\lambda)=\int_{-\pi}^{\pi} f(\lambda)d\lambda$ is a spectral distribution function (cumulative) and $f(\lambda)$,
$\pi \leq \lambda \leq \pi$, is the spectral density of Z_t. Eqn. (3.9) is referred to as the zero-crossing
spectral representation [29] of a random process. It shows that the normalized
expected zero-crossings per unit time, $\pi E\{D\}/(N - 1)$, which is the zero-crossing
rate of the random process, Z_t, may be represented in terms of the angular (weighted)
spectral content of Z_t. Therefore, within a subband, if a certain frequency, ω_0,
becomes large or dominant within a frequency band $\omega \in [0, \pi]$, that is, has more
power (weights) than in other bands, then it is seen from Eqn. (3.9)

$$\frac{\pi E\{D\}}{(N - 1)} \approx \omega_0 \tag{3.10}$$

which states that this dominant frequency may be represented by a corresponding
average zero-crossing rate. For the case when only one frequency, ω_0, is present for
$\omega \in [0, \pi]$,

$$F(\omega+) - F(\omega-) \quad > \quad 0, \qquad \omega = \omega_0$$
$$= \quad 0, \qquad \omega \neq \omega_0 \qquad (3.11)$$

from which is obtained the equality

$$\frac{\pi E\{D\}}{(N-1)} = \omega_0. \qquad (3.12)$$

Replacing $E\{D\}$ by D, it is seen that a dominant frequency in a frequency band can be readily detected by zero-crossings of the random process within a time interval of length N from the expression $\pi D/(N-1)$. This is referred to as the dominant frequency principle [29]. A dominant frequency in a subband has more power than others, which makes the zero-crossing algorithms more robust in noisy conditions [7]. The resolution of the subband dominant frequency estimates decreases rapidly at higher frequencies. One method to circumvent this is by up-sampling the high frequency subband signals using frequency-dependent interpolation factors [8].

The ZCPA auditory model as proposed by Kim *et al.* [7] is based on the dominant frequency principle and duplicates the functionalities of the peripheral auditory system in considerable detail. It is an enhancement of the ensemble interval histogram (EIH) model [19] where it replaces the multiple levels of the EIH with a single zero level for feature extraction. Fig. 3.2 shows the schematic block diagram of the ZCPA auditory model. It utilizes a zero-crossing detector to estimate the dominant frequencies within a filterbank channel. Each speech frame is represented by a frequency histogram obtained from the inverse of the interval between two successive upward going zero-crossings from all the channel outputs within the frame. The intensity information is measured by a peak detector from the peak amplitudes between each zero-crossing interval. In this experiment, a computationally efficient version of the ZCPA with fewer number of filters to reduce the computational cost and suitable for hardware implementation was implemented.

Let $m_s(n)$ be the clean speech signal which is corrupted with bandlimited white Gaussian noise, $\nu(n)$, with a rectangular power spectrum of bandwidth, W, such that the noisy speech is given by

$$s(n) = m_s(n) + \nu(n) \qquad (3.13)$$

The input noisy speech $s(n)$ may be expressed as

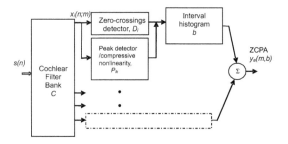

FIGURE 3.2: Schematic of the ZCPA auditory model.

$$s(n) = \sum_{l=0}^{M-1} A_l \cos(\omega_l n + \theta_l) + A_\nu \nu(n) \tag{3.14}$$

where $l = 0, 1, \ldots, M-1$ corresponds to the dominant frequency components within
the signal. Let the input noisy signal, $s(n)$, be filtered by a bank of bandpass cochlear
filters with bandwidth, B_i, where $i = 1, \ldots, C$ corresponds to the number of channels
in the filterbank. Then the output at the i-th bandpass filter, assuming that each
sinusoidal component of $s(n)$ is separated by the filterbank with only w_l in filter i,
may be expressed in discrete form as,

$$x_i(n) = A_l \cos(\omega_l n + \theta_l) + A_\nu \nu_i(n). \tag{3.15}$$

If $w(n)$ is a window of finite length, then the value of $x_i(n)$ at the frame index, m,
is given by $x_i(n; m) = x_i(n) w_i(m - n)$, $i = 1, \ldots, C$, where C is the number of filter
channels.

If D_i is the number of upward-going zero-crossings at the i-th filter in a windowed
frame $x_i(n; m)$ and P_{ik} is the peak amplitude between the k-th and the $(k+1)$-th
zero-crossing in $x_i(n; m)$, then the output $y(m, b)$ of the ZCPA at time m is given
in [7] as

$$y(m, b) = \sum_{i=1}^{C} \sum_{k=1}^{D_i-1} \log(P_{ik} + 1)\delta_{bj_k}, \qquad 1 \le b < R \tag{3.16}$$

where R is the number of frequency bins, and δ_{bj} is the Kronecker delta, such that

$$
\begin{aligned}
\delta_{bj_k} &= 1 \quad \text{if} \quad b = j_k \\
&= 0 \quad \text{elsewhere.}
\end{aligned}
\tag{3.17}
$$

From each channel output, the index of frequency bin, j_k, is computed by taking the inverse of the time interval between the k-th and $(k+1)$-th zero-crossings, for $k=1,\ldots,D_l$-1. For each k within a frame, the value of the frequency histogram at the frequency bin j_k is increased by logarithm of the peak value within the zero-crossing interval, P_{lk}. The histogram across all channels are combined to obtain the interval histogram for that frame. The interval histograms for all frames are combined to form the ensemble interval histogram output of the ZCPA, $y_a(m,b)$.

In the ZCPA, a synchronous neural firing ("spike") is simulated as the upward going zero-crossing event of the input stimulus [7]. In the temporal-rate representation of the auditory nerve coding, the auditory nerve activity occurs in synchrony with the stimulus periods. At low CFs the dominant frequencies are resolved with high precision and the neural discharges of the auditory nerve fibres are phase locked to the formants. The synchrony information is extracted from the ensemble interval histogram, where the expected value of the zero-crossing count, D, is related to its spectral representation given by the Eqn. (3.9). However, at high CFs, frequency resolution is poor due to the wider bandwidths, and the phase-locking of the discharges is greatly reduced. In such case, the instantaneous rate of firing conveys temporal information with fine time resolution.

If D is considered a random variable, then the variance of the time interval perturbation (TIP) between two adjacent zero-crossings of a noisy signal, $s(t)$, corrupted by white noise, $\nu(t)$, at a particular channel zero-crossing detector is given in [7] as

$$\sigma_l^2 = \frac{2A_\nu^2 B_l/W}{(\omega_l A_l)^2} \qquad (3.18)$$

where B_l is the bandwidth of the cochlear filter corresponding to the frequency component l, $l \in [0, M]$, at which the variance is measured, and W is the rectangular power spectrum bandwidth of the white Gaussian noise, $\nu(t)$. It is observed from Eqn. (3.18) that the zero-crossing variance is inversely proportional to the frequency components present in the input speech.

3.1.1 The temporal adaptation model

In the base ZCPA temporal auditory model the firings of the auditory nerve are simulated by the positive going zero-crossings of the input stimulus which are not functions of synaptic processing which take place in the peripheral auditory system. Moreover, dynamic or broadband transient properties related to synaptic processing are not implemented. Therefore, it is expected that the ASR performance may be

further improved by integrating synaptic processing, such as the synaptic adaptation, in the feature extraction process. Substituting Eqn. (3.12) for the dominant
frequency into Eqn. (3.18), we obtain a relation of the zero-crossing variance, σ_l^2, in
terms of the zero-crossings, D_l, for the dominant frequencies in the i-th channel as

$$\sigma_l^2 = \frac{2A_\nu^2 B_l / W (N-1)^2}{\pi^2 E\{D_l\}^2 A_l^2}. \tag{3.19}$$

It is seen from Eqn. (3.19) that the variance of the zero-crossing perturbations
is reduced more for higher zero-crossing rates corresponding to the high frequency
components than for the lower frequency components. A high-pass filter may be
used to further enhance the high frequency components and to take advantage of
the variance reduction property of the ZCPA at higher frequencies.

A reduction in variance of the zero-crossing perturbation may increase the precision of the unbiased estimates of the HMM model parameters that can be obtained
by an iterative algorithm such as EM. This can be seen from the following analysis.

If θ is the true value and $\hat{\theta}_N$ is an estimate of the parameter of a random variable,
where N is the number of observations, then for an unbiased estimate, $\theta = E\{\hat{\theta}_N\}$.
In order for the estimate of a HMM parameter to converge to its true value, it is
necessary that the variance of the estimates go to zero as the number of observations
go to infinity

$$\lim_{N \to \infty} \text{Var}\{\hat{\theta}_N\} = \lim_{N \to \infty} E\{|\hat{\theta}_N - E\{\hat{\theta}_N\}|^2 = 0. \tag{3.20}$$

For an unbiased estimate of θ, it is shown using the Tchebycheff inequality [92] that
for any sufficiently small $\epsilon > 0$,

$$\Pr\{|\hat{\theta}_N - \theta| \geq \epsilon\} \leq \frac{\text{Var}\{\hat{\theta}_N\}}{\epsilon^2}. \tag{3.21}$$

Therefore, if the variance is sufficiently small, then the probability that the difference
between the true value and the estimated value is greater than ϵ will also be smaller.
In the case of noisy speech the covariance matrices are always decorrelated so that
the diagonal elements are the individual variances of the random process.

To implement a simplified adaption scheme suitable for ASR applications, we
defined the first order high-pass infinite impulse response (IIR) filter function as

$$H_a(z) = \frac{10\tau f_r (1 - z^{-1})}{(10\tau f_r + 0.05) + (10\tau f_r - 0.05)z^{-1}} \tag{3.22}$$

where τ is the synaptic adaptation time constant in seconds and f_r is the frame rate
equal to 100 Hz [93]. Fig. 3.3 shows the rapid and the short-term synaptic adapta-

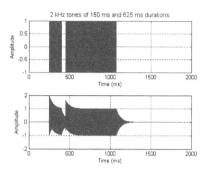

FIGURE 3.3: Rapid and short-term synaptic adaptation effects to 2 kHz tone bursts of 150 and 625 ms durations with a time constant $\tau=40$ ms using the transfer function $H_a(z)$ (Eqn (3.22).

tion responses using Eqn. (3.22) with a time constant $\tau=40$ ms to a 2 kHz tone of 150 ms duration followed by a tone of 625 ms duration. The initial onset transient and the decay due to the temporal synaptic adaptation process are observed in the figure.

In Fig. 3.3 effects similar to forward masking are also observed, e.g., the second tone burst is reduced in magnitude due to the presence of the first tone burst preceding it when the time duration separating the two tones is less than the time constant of the filter. If the time constant is large and the frame size is relatively small compared to the synaptic adaptation time constant, the forward masking effect will depend on the previous frames [94]. However, for the temporal synaptic adaptation we did not consider the effects of the previous frames. Speech perception is more sensitive to rapid variations and dynamic changes than the steady state characteristics. In synaptic adaptation, one of the objectives is to enhance the speech onsets, which occur within the initial 15 ms of the input stimulus. This effect may be extracted from a reasonable frame size, usually 30-120 ms, which are employed for feature extraction in temporal auditory models. This effect of synaptic adaptation on onsets is more prominent than the forward masking effect across the frames. Moreover, in the HMM a state corresponds to an acoustically stable region or a number of continuous number of frames with stationary observation vectors. Onset region detection techniques based on a hidden Markov model (HMM) utilize localization of onset regions as temporal sequences within a frame which are then combined together to be fed into a HMM. Frames containing onsets regions in the speech can be detected directly from the state transition sequence by the HMM. Moreover, by using a frame advance of 10 ms a frame-based method can still capture some of the rapid synaptic adaptation effects, depending on the time alignment of the frame with respect to the onset time. In conventional feature extraction by spectral methods, detection of CVC transitions in speech which are typically dynamic in nature like synaptic adaptation with higher durations typically 80-100ms, follow a similar detection procedure by a frame-based method utilizing 25 ms frame windows. Although the effects are much enhanced by incorporating delta-delta features.

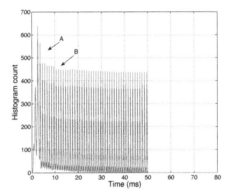

FIGURE 3.4: PSTH obtained from the ZCPA with an synaptic adaptation filter with a time constant $\tau=40$ ms, showing the effects of rapid synaptic adaptation (pointer 'A') and short-term synaptic adaptation (pointer 'B'). The PSTH was taken with 70 presentations of a 1 kHz tone input of 50 ms duration in 400 bins with a bin width of 0.2 ms.

It has been reported that forward (post-masking) and simultaneous masking can last up to 200 ms, whereas backward masking (pre-masking) are of much shorter duration, typically 10-30 ms [27]. Longer time constants are important for speech processing which may give better recognition performance. The best time constant for ASR lies between 200 and 300 ms [27]. A time constant of 250 ms was used in all our experiments utilizing temporal synaptic adaptation with a cutoff frequency of 0.636 Hz which is well below the modulation spectrum of 1-16 Hz. The modulation frequencies are regarded important for speech intelligibility [71] and the cutoff of 0.636 Hz of the high-pass filter would not interfere with the modulation frequencies. The synaptic adaptation filtering was implemented by summing the high-pass synaptic adaptation filter output with the original FIR filter output for each channel of the filterbank.

Fig. 3.4 shows the post-stimulus time histogram (PSTH) obtained from the

ZCPA auditory model applied with a synaptic adaptation filter with a time constant τ=40 ms. The input was a 1 kHz tone of 50 ms duration. The PSTH was taken with 70 presentations of the input and collected in 400 bins with 0.2 ms bin-width. It was observed that at the onset transient, there was a rapid increase in the histogram count, followed by a rapid decrease (due to the rapid synaptic adaptation) within about 3 ms (pointer 'A'), followed by short-term synaptic adaptation (pointer 'B') and a nearly constant steady state response. It is observed in Fig. 3.4 that effects similar to the rapid synaptic adaptation may be implemented in temporal processing of speech such as in the ZCPA, not otherwise possible in spectral implementation. The Figs. 3.3 and 3.4 have been implemented with pure tone bursts to demonstrate the effects of synaptic adaptation using Eqn. (3.22). Speech signals are, however, more complex than simple tone bursts, and the time constant of 40 ms may not be appropriate for speech recognition purposes.

Westerman *et al.* [31] measured the time constants in the Mongolian Gerbil to be a few milliseconds for rapid synaptic adaptation and roughly 40-60 ms for short-term synaptic adaptation. Spoor and Eggermont [89] found that synaptic adaptation time constants in humans might differ considerably from values measured in animals and estimated that human time constants of recovery from synaptic adaptation are about a factor of four longer compared to gerbil data. It has been reported that forward (post-masking) and simultaneous masking can last up to 200 ms whereas backward masking (pre-masking) are of much shorter duration, typically 10-30 ms [95]. This was verified by Holmberg by a series of experiments, evaluating word error rates for a range of synaptic adaptation time constants. It was found that for the case of clean training with Wiener filtering, the best time constant is 80 ms, but for the cases with adaptation and multi-condition training, it was found that the best time constant lies between 200 ms and 300 ms. Therefore an synaptic adaptation time constant of 240 ms was chosen by Holmberg in all experiments. This time constant is also motivated by the notion that higher levels in the auditory pathway, with presumably longer adaptation time constants than the auditory nerve, contribute to the temporal processing of speech. The 240-ms time constant is consistent with other auditory models that have been used as ASR front ends [22] and with the highpass corner frequency employed in RASTA [71] processing. Motivated by these experimental results, we have chosen a time constant of 250 ms as the adaptation time constant in all our experiments utilizing temporal synaptic adaptation. A time constant of 250 ms gives a cutoff frequency of 0.636 Hz which is well below the modulation spectrum of 1-16 Hz. The modulation frequencies are regarded important for speech intelligibility [71] and the cutoff of 0.636 Hz of the high-pass filter would not interfere

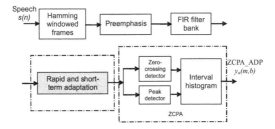

FIGURE 3.5: Schematic of the ZCPA with synaptic adaptation (ZCPA_ADP). The arrow at the right of the top row from each channel of the filterbank connects to the arrow at the left of the bottom row and each output in the filterbank is processed separately and combined into an ensemble interval histogram.

with the modulation frequencies. The synaptic adaptation filtering was implemented by summing the high-pass synaptic adaptation filter output with the original FIR filter output for each channel of the filterbank.

3.1.2 Feature extraction with synaptic adaptation

Fig. 3.5 shows the ZCPA with the temporal synaptic adaptation used as a pre-processing front-end. The synaptic adaptation as implemented in Eqn. (3.22) was introduced in each filter channel with a time constant $\tau=250$ ms. The input speech was first normalized between ±1 to reduce the effects of loudness. Normalization also improves noise performance by enhancing spectral contrast. Speech frames were pre-emphasized to model the outer and the middle ear functionalities that approximate the unequal sensitivity of human hearing at different frequencies. A pre-emphasis coefficient of -0.97 was used. It was then processed by a bank of 16 finite impulse response (FIR) filters of order 70 with the characteristic frequencies (CF) uniformly spaced on the equivalent rectangular bandwidth (ERB) scale [42] between 10-3500 Hz. The frequency response of the perceptual FIR filterbank is shown in Fig. 3.6.

Our choice of FIR filters for frequency processing was motivated by the fact that FIR filters consistently performed better than carefully designed cochlear filters when applied in ASR applications [7]. In most applications, real-time operations through hardware implementation is often the main goal. All digital signal processors (DSPs) available have architectures in powers of two. That is, 2^n, where n is the number of available bits, is the the number of ways the bits may represent in a

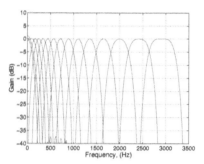

FIGURE 3.6: Frequency response of the perceptual FIR filterbank consisting of 16 FIR filters each of order 70 with CF spacing at the ERB scale.

digital system. This is particularly suited to FIR filtering in hardware. FIR filters are always stable, have exactly linear phase, and are simple to implement. However, FIR filters, particularly of higher orders, may also introduce substantial lag or time delay. Since in the ZCPA, the output are extracted as a frequency histogram counts, the time delay effects on the output are not considered [7].

To reduce computational time of the ZCPA, we implemented the simplified model with fewer number of filters with some optimization of the parameters and feature extraction algorithm. However, fewer number of filters result in greater frequency overlap of adjacent frequency channels and also introduces a histogram bias in the extracted features. This may result in reduced frequency resolution or a frequency bias in the ZCPA histogram at the cost of faster and a more efficient processings. The biasing in the histogram which may arise due to the use of fewer number of filters may cause two effects. firstly, this will affect the overlapping of the adjacent filters, especially for the higher frequency filters, where the overlap is higher due to the wider bandwidths. This will bias the high frequency histogram counts more than the low frequency histogram counts. Secondly, modelling inaccuracies may result in the extracted features due to this overlap. Gajic et al. [96] made a study of the ASR performance obtained by several different choices of filter bandwidths and number of histogram bins and has quantified the results. It was observed that the choice of filter bandwidths had a significant influence on the ASR performance of ZCPA features, while it was not very sensitive to the particular choice of the number of histogram bins. However, the influence of increased number of filters was tested by evaluating ZCPA features based on 16 and 20 filters, and no significant performance

71

difference was observed [96]. Therefore, although there is a bias in the histogram due to reduced number of filters used, the effects on ASR performance is not very significant.

In the ZCPA, the width of the derivative windows in units of time over which the features are extracted are functions of the centre frequencies of the filterbank. In temporal processing it is usually required to use large window sizes especially for lower frequency channels [96] to capture about 10 periods of the signal for the accumulation of temporal information. Although longer time windows give better parameter estimates, the window size should not be too large to violate the stationarity assumption in the short-term processing method. Considering these, the largest window size was limited to 80 ms corresponding to the lowest frequency channel. In the filter subbands, the derivative windows were made proportional to the inverse of the channel centre frequencies. Accordingly, the zero-crossing intervals were collected over a derivative window length of $10/f_k$ for lower frequencies and $60/f_k$ for higher frequencies, where f_k is the filter centre frequency [96]. For example, for a f_k (which is the same as CF) of 125 Hz, the window length would be 80 ms, whereas for a f_k of 2000 Hz, the window length would be 30 ms. In each filter output, inverse of zero-crossing intervals were collected in 26 frequency bins uniformly spaced between 10-4000 Hz on the ERB scale. The interval histogram was weighted by the logarithm of the peak value within two successive zero-crossings. One ZCPA frame was obtained every 10 ms. The histogram was normalized between 0 and +1 with respect to the maximum value. Normalising the histogram produces two effects Firstly, it limits the dynamic range of the histogram within 0 and 1 and secondly, the compression due to normalising reduces the effects of any biasing in the histogram. Thirteen cepstra were generated and retained from each speech frame.

Figs. 3.7 (a) and (b) show the spectrogram of the 35-th frame for the male CV utterance /ba/ without adaption and with temporal synaptic adaptation τ=250 ms, respectively [97]. The processing started at the 35th frame of the whole utterance /ba/ and ended after the processing of that frame. A single frame was taken to study the time-frequency characteristics within this frame with greater resolution and for comparison for the two cases with synaptic adaptation and without synaptic adaptation. It is assumed that the steady-state vowel /a/ and the corresponding formants are captured in this frame in addition to the high frequency speech segments in that frame. It is observed in (b) that the high frequency segments are enhanced by the application of the temporal synaptic adaptation strategy. However, the formants have little effects due to the synaptic adaptation.

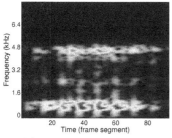

(a) Without synaptic adaptation

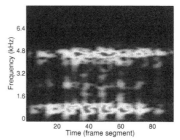

(b) With synaptic adaptation

FIGURE 3.7: Spectrogram of the ZCPA for the 35-th frame of the male utterance /ba/ in clean condition (a) without synaptic adaptation, and (b) with synaptic adaptation with a time constant τ=250 ms showing enhanced high frequency segments. (In gray scale, darker shade implies lower intensity and lighter shade indicates higher intensity).

3.2 CV (consonant-vowel) discrimination by temporal auditory adaptation

ASR relies to a larger extent on the correct classification of vowels than consonants [28]. Temporal auditory models usually employ longer window lengths which are particularly sensitive to non-stationary consonant-vowel (CV) transitions which are poorly represented in such models. ZCPA with temporal synaptic adaptation was further tested in a CV discrimination task using the UCLA-SPAPL CV speech corpus. The UCLA-SPAPL CV is an extensive database of isolated CV utterances.

Each word is uttered in 8 tokens by each of the four speakers (2 males and 2 females) using 18 consonants, each consonant in three vowel contexts, thus constituting a total of 1728 utterances. For example, corresponding to the plosive /b/, the utterances are /ba/, /bee/ and /boo/, corresponding to the three vowels /a/,/ee/ and /oo/. The CV combinations were digitally recorded at a sampling rate of 16 kHz, and are shown labeled in ASCII in Table 3.1.

In our experiment, two analyses were performed to study the effects on speech perception with the simplified synaptic adaptation scheme: articulatory discrimination in CV utterances and vowel clustering analysis by scatter plots.

73

TABLE 3.1: Consonants and the corresponding vowel tokens in the UCLA-SPAPL CV corpus.

Articulation	Consonants	Vowel contexts
Voiced plosives	/b/,/d/,/g/	
Aspirated plosives	/p/,/t/,/k/	/a/ as in "bad"
Nasals	/m/,/n/	/ee/ as in "beat"
Fricatives	/s/,/Z/,/sh/,/SH/ /f/,/V/,/th/,/TH/	/oo/ as in "boot"
Affricates	/j/,/ch/	

3.2.1 A measure of CV discrimination by synaptic adaptation

The IIR adaptation filtering primarily enhances the high frequency components associated with frication and voice onsets. Since most speech energies are concentrated more in the lower frequency regions, the contributions to the high frequency histograms were mainly due to the increased zero-crossings counts instead of the contributions of the interval peak values. Hence, a quantitative measure of the effects of the high pass synaptic adaptation filtering on speech perception could be obtained by summing the histogram counts at the output of the channels with CFs equal or higher than 1340 Hz. The choice of this frequency is rather arbitrary and was partly based on consideration that it should be higher than the first formant, which is the dominant formant of the common vowels carrying the maximum energy compared with the higher formants. Another consideration for not using a higher frequency than the chosen one is that at higher frequencies, other high frequency artefacts of speech such as consonants (voiced and unvoiced plosives and fricatives) play a more important role, and may compromise the experimental results.

The vowel discrimination with synaptic adaptation was measured as the percentage increase in the histogram counts with synaptic adaptation (ZC_{adp}) over the histogram counts without synaptic adaptation (ZC), as obtained using the formula

$$\frac{\sum_f ZC_{adp}(f) - \sum_f ZC(f)}{\sum_f ZC(f)} \times 100\%, \qquad f \geq 1340 \quad \text{Hz}. \qquad (3.23)$$

The consonant types /b/,/d/,/g/,/p/,/t/,/k/,/m/,/n/,/s/,/SH/,/j/,/ch/ in the three vowel contexts /a/,/ee/and /oo/ were used in the experiment. The results are summarized in Fig. 3.8. It is observed that the largest percentage increase in the histogram counts are recorded for the vowel /ee/ in all consonant types (14.16 % for /SHee/), followed by the vowel /oo/ in fricatives (11.49 %) and affricates (9.19

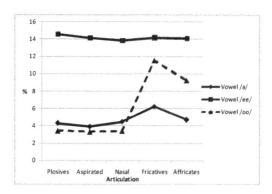

FIGURE 3.8: Effects of temporal synaptic adaptation on vowel perception as percentage increase in the high frequency histogram counts (y-axis) for ZCPA with adaptation, (ZC$_{adp}$), over the base ZCPA, (ZC).

%). This may be due to the higher second formant for the vowel /ee/ (1730 Hz) compared to vowel /a/ (1090 Hz) and vowel /oo/ (870 Hz).

3.2.2   Vowel clustering with synaptic adaptation

Clustering is the classification of similar objects into different groups. Partitioned clustering segregates data so that the degree of association is strong between members of the same cluster and weak between members of different clusters. The mixture likelihood approach to clustering, which is a class of unsupervised learning, is based on maximization of the likelihood estimation for finite mixture models utilizing the EM algorithm. The method utilizes a iterative solutions of the likelihood equations for obtaining the estimate of the unknown parameters, such as the mean and variances associated with each mixture model [98]. The estimate of the posterior probabilities can then be formed for each observation data to give a probabilistic clustering, depending on the highest estimated posterior probabilities (Sec. 2.7.2).

A more simple measure of unsupervised clustering may be obtained by the linear discriminant analysis [99]. In our experiment, a qualitative measure of clustering with synaptic adaptation of the three vowel classes in the UCLA-SPAPL database was obtained by grouped scatter plots of the CV utterances, expressed as an utterance matrix, each row corresponding to an utterance represented by a m-cepstrum vector in the m-dimensional space. In two dimensional linear discrimination, any two fixed dimensions of the utterance matrix may be plotted as the x and y coordinates,

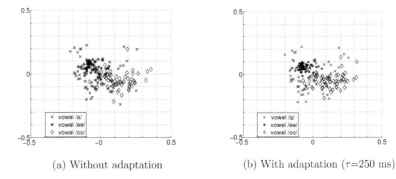

(a) Without adaptation   (b) With adaptation (τ=250 ms)

FIGURE 3.9: Two dimensional scatter plots of the utterance matrix obtained from the voiced and unvoiced plosives /b/,/d/,/g/ and /p/,/t/,/k/, respectively, grouped by the three vowel contexts /a/, /ee/, /oo/ for (a) base ZCPA and (b) ZCPA with synaptic adaptation (τ=250 ms) in clean condition.

with the means associated with each class as the centroid. The discrimination among the classes is essentially reflected in the extent of the correlation between these two components of the feature vector set, as exemplified by scattering from the individual class centroids. For our experiments, we selected the voiced plosives /b/,/d/,/g/ and the unvoiced aspirated plosives /p/,/t/,/k/ in three vowel contexts /a/, /ee/ and /oo/, for a total of 18 utterances. Only the plosives were selected considering that these have the most prominent onset regions and hence, are important and significant for the synaptic adaptation process. The 18 CVs were uttered by 2 male and 2 female speakers, each speaker uttering the same CV in 5 tokens, thus constituting a total of 360 CV utterances. Features were extracted from each utterance from a ZCPA interval histogram at the rate of 13 cepstra per frame at a frame rate of 10 ms. Each utterance was collapsed into a single vector by taking the mean of all the frames, thus converting the 360 utterances into a single 360x13 utterance matrix. The two dimensional grouped scatter plots of the utterances matrix is shown in Fig. 3.9 without synaptic adaptation in (a) and with synaptic adaptation in (b). The increased clustering of the vowel /ee/ having a higher second formant (1730 Hz) is observed in (b) which is consistent with the results of Sec. 3.2.1. However, the other vowels /aa/ and /oo/ having lower second formants (1090 Hz for /a/ and 870 Hz for /oo/) appears to be just as scattered. This further emphasizes the role of synaptic adaptation on high frequency articulation and dynamic behaviours of speech, particularly on the second and higher formants. Improved speech recognition performance may result from this effect.

3.2.3 Linear discriminant analysis of synaptic adaptation by class separability measure

Measuring the discrimination effectiveness of feature vectors is an important method of determining robustness which aids in the selection of optimal features for a given dimensionality. Class separability measures the discrimination effectiveness of feature vectors. Measuring the discrimination effectiveness may assist and can be used to evaluate a particular algorithm for feature extraction process. In Sec. 3.2.2, a qualitative measure of clustering with synaptic adaptation was presented. A quantitative measure of the separability, or divergence among the three vowel classes may be determined by one of several methods such as Kulback-Liebler distance measure between density functions [100], and the Brattacharayya distance [101]. However, determination of these separability measures are feasible only when a Gaussian assumption of the distribution of the class members is valid.

A simpler and more general quantitative criterion may be obtained from the scatter characteristics of the feature vector samples in the m-dimensional space using the linear discriminant analysis approach based on the within-class and between-class scatter matrices [78]. It utilizes a linear transformation which maps the feature space into a lower dimensional space such that a class separability criterion is optimized.

Let $\{\mathbf{x}_n\}_{n=1}^{K}$ be a set of feature vectors each of dimension $K=13$, $N=60$ is the number of observations in each class w_l, $l \in L$, where $L=3$ is the total number of classes. Defining the mean vector, $\boldsymbol{\mu}_l$ and the covariance matrix, Σ_l, for each class l as

$$\boldsymbol{\mu}_l = \frac{1}{N}\sum_{n=1}^{N}\mathbf{x}_n, \qquad \mathbf{x}_n \in w_l. \tag{3.24}$$

$$\Sigma_l = \frac{1}{N}\sum_{n=1}^{N}(\mathbf{x}_n - \boldsymbol{\mu}_l)(\mathbf{x}_n - \boldsymbol{\mu}_l)^{T}, \tag{3.25}$$

the separability criterion may be defined as a function of two scatter matrices: the within-class scatter matrix, S_w, and the between-class scatter matrix, S_b. The within class scatter matrix, S_w, represents scatter around their class mean vector $\boldsymbol{\mu}_l$, while the between class scatter matrix, S_b, represents the scatter of the class means around the total mean $\boldsymbol{\mu}$. The two matrices, S_w and S_b, are expressed as

$$S_w = \frac{1}{(N-1)}\sum_{l=1}^{L}\sum_{\mathbf{x}\in w_l}(\mathbf{x}_n - \boldsymbol{\mu}_l)(\mathbf{x}_n - \boldsymbol{\mu}_l)^{T} \tag{3.26}$$

$$S_b = \frac{1}{N} \sum_{l=1}^{L} N_l (\boldsymbol{\mu}_l - \boldsymbol{\mu})(\boldsymbol{\mu}_l - \boldsymbol{\mu})^T. \tag{3.27}$$

From the above matrices two measures of separability, J_1 and J_2, among the three vowel classes is obtained as

$$J_1 = \mathrm{tr}(S_w^{-1} S_b) \tag{3.28}$$

$$J_2 = \frac{\det(S_b + S_w)}{\det(S_w)}. \tag{3.29}$$

It is seen that S_w is the covariance matrix of individual classes summed over all classes and its traces are sum of variances. J_1 takes large values when samples in the l-dimensional space are well clustered around their mean, within each class, and the clusters of the different classes are well separated. It has the advantage of being invariant under linear transformations [78]. J_2 gives an alternate criterion with determinants used in place of traces. This is justified for scatter matrices that are symmetric positive definite and thus their eigenvalues are positive and the trace is equivalent to the sum of the eigenvalues [78]. Hence, these measures are useful benchmarks of separability for better feature extraction and for speech recognition purposes.

For our experiment, we utilized the three voiced plosives /b/,/d,/g/, each in three vowel contexts, /a/,/ee/,/oo/. The subset utilized only voiced plosives for the clustering experiment since these have the most prominent onset regions and hence, are important and significant for the synaptic adaptation process. The 9 CVs were uttered by 4 speakers in 5 different tokens, for a total of 180 CV utterances, with 60 utterances in each of the three vowel classes. For $N=60$ and $L=3$, the computed measures of separability, J_1 and J_2, are shown in Table 3.2. It is observed that the ZCPA_ADP showed higher class separability of the three vowel classes than the base ZCPA.

TABLE 3.2: Measures of separability, J_1 and J_2, among the three vowel classes /a/, /ee/ and /oo/ based on LDA of the within-class and between-class scatter matrices for the base ZCPA and ZCPA with synaptic adaptation.

	J_1	J_2
Base ZCPA	21.98	112.50
ZCPA_ADP	24.49	120.49

3.3 HMM word recognition with synaptic adaptation

Speaker independent isolated digits from the TIDIGITS speech corpus were used for recognition experiments. There were 55 male speakers in the training set, each speaker with 2 utterances of the digits 1-9, 'oh' and 'zero' (total 1210 utterances in the training set) and and a separate set of 31 male speakers in the test set each speaker with 2 utterances of the digits 1-9, 'oh' and 'zero' (total 682 utterances in the test set). Continuous Gaussian density HMM with 15 states per digit, 5 mixture components per state with diagonal covariances were used to define each model. A 3-state silence model was inserted at the beginning of each utterance. The Baum-Welch re-estimation using a flat-start scheme and 12 estimation iterations was used for training under clean condition. Test speech was corrupted with 4 different types of additive noise from the NOISEX 92 database: Gaussian white noise, factory noise, babble noise and Volvo noise. In the testing phase, a left-to-right Viterbi recognizer using word-level lattice network, a dictionary and a token passing algorithm with dynamic programming was used for decoding. Recognition results are shown in Table 3.3.

TABLE 3.3: Recognition rates (%) of the ZCPA and ZCPA with synaptic adaptation (ZCPA_ADP) for isolated digits (TIDIGITS corpus) with training and testing with male utterances in four types of additive noise.

SNR (dB)	White		Factory		Babble		Volvo	
	ZCPA	ZCPA _ADP	ZCPA	ZCPA _ADP	ZCPA	ZCPA _ADP	ZCPA	ZCPA _ADP
Clean	95.4	95.4						
40	90.9	95.4	95.4	90.9	90.9	90.9	95.4	95.4
30	81.8	90.9	86.3	81.8	86.3	86.3	90.9	90.9
15	77.3	72.7	81.8	77.3	72.7	72.7	86.3	86.3
10	68.2	59.1	68.8	72.7	63.6	59.1	86.3	81.8
5	50.0	31.8	50.0	68.8	31.8	54.5	81.8	77.3

In speech recognition tests, it is important to perform a statistical significance test to determine whether the differences in error-rates between two algorithms tested on the same data set are statistically significant. The McNemar's test [102] can be used for this purpose, which requires the errors made by an algorithm to be independent events and is applicable for isolated word recognition algorithms, that is, if the uttered words are not context dependent or dependent on some language model. The recognition results in Table 3.3 were further tested using the McNemar's test for the 682 test utterances and it was found that the differences in recognition results due to the two algorithms, (base ZCPA and ZCPA with synaptic adaptation) were statistically significant whenever a difference in recognition result is observed. The McNemar's significance test procedures are stated in Appendix A.

It is observed from Table 3.3 that the performance of the base ZCPA (without synaptic adaptation) follows the same trend as given by Kim *et al.* in [7]. That is, the best performance is achieved in Volvo noise, followed by factory noise, white noise and the babble noise. Thus the response depends on the type of noise, with better performance in non-Gaussian real-world noise environment. This is due to the fact that real-world noise contain significant low frequency components and are characterized by the absence of high frequency noise perturbations which usually corrupt the lower intensity high frequency articulatory cues. Moreover due to the dominant frequency principle of the ZCPA, the formants are preserved in presence of low frequency noise perturbations.

For the ZCPA with synaptic adaptation, no improvement is observed in clean condition over the base ZCPA. Improvements are observed in low noise conditions for the case of white noise and in high noise conditions for the case of factory noise. In presence of noise, the same trend is detected, with the best performance with Volvo noise and the worst performance with babble noise. However, compared to the base ZCPA, improvements with synaptic adaptation is observed only with white Gaussian noise at high SNRs, with degradation at low SNRs. With factory noise, a reciprocal effect, that is, a degradation at high SNRs with improvement at low SNRs, is observed. In both babble and Volvo noise, there was either equal or degraded performance, particularly at low SNRs. In general, temporal synaptic adaptation performed better in stationary Gaussian white noise except at very low SNRs, but performed poorly in non-Gaussian real-world noise. The improvements may be observed from Eqn. (3.18) and Eqn. (3.19), in which it is seen that the zero-crossing variance is reduced for the higher frequency noise components. In white Gaussian noise, the noise perturbations exist at all frequency components, that is, it has a flat spectral density, with significant high frequency components than the real world or man-made noise, such as factory, babble and Volvo noise. The high pass synaptic adaptation filter further enhances the high frequency components, which further reduces the variance.

For non-stationary factory noise, which contain significant low frequency components, the effects with high-pass synaptic adaptation filtering are expected to be reciprocal to that of stationary white noise, with worse performance at high SNRs, and better performance at low SNRs, as observed from Table 3.3. The recognition performance for babble and Volvo noise is similar to factory noise with no improvements or degradations mainly due to the presence of significant low frequency components.

Similar dependence of auditory synaptic adaptation on high frequency articulation was observed in Sec. 3.2.1 where it was observed that the largest percentage increase in the histogram counts were recorded for the vowel with higher frequency second formants (for the vowel /ee/ (1730 Hz) compared to vowel /a/ (1090 Hz) and vowel /oo/ (870 Hz)).

It is reported in [7], that although the base ZCPA performs better than MFCC in noisy conditions, the relative performance improvement for the ZCPA in noisy conditions over the MFCC is greater in white Gaussian noise than in real-world noise, such as factory noise. It is stated therein that the reason for this is not clear. One reason for this may be that in the ZCPA, the zero-crossing variance is reduced more for the high frequency white noise components than the lower frequency real-world noise (factory, car, and babble noise) components. Therefore, although overall the ZCPA performs better in real-world noise than in white noise, the relative improvement in performance compared to MFCC is higher in white noise than in non-Gaussian real-world noise.

However, it is observed from Table 3.3 that at very low SNRs, there is a degradation in recognition rates in white noise. This is due to three reasons. Firstly, if the unit sample response, $h(n)$, of the FIR filter is finite in length and zero outside the interval $[0,N\text{-}1]$, and $\boldsymbol{\nu}=[\nu(0), \nu(1), \ldots, \nu(N-1)]$ is the uncorrelated input noise with zero mean, then the variance (power) of the noise at the output of the i-th FIR filter can be expressed as

$$\sigma_{\nu_i}^2 = E\left\{|\nu_i(n)|^2\right\} = \mathbf{h}^H \mathbf{R}_\nu \mathbf{h} = \mathbf{h}^H \mathbf{C}_\nu \mathbf{h} = \mathbf{C}_\nu \mathbf{h}^2 \qquad (3.30)$$

where $\mathbf{h}$ is the vector of filter coefficients, $\mathbf{R}_\nu$ is noise autocorrelation matrix and $\mathbf{C}_\nu$ is autocovariance matrix whose diagonals are simply the variances of the input noise process. Thus, we see that the variance of the noise $\nu(n)$ at the output of the FIR filter is obtained by multiplying the input noise variance by the square of the FIR filter coefficient vector, $\mathbf{h}$. Hence, for a large order FIR filter, output noise variances may increase substantially.

Secondly, it is observed from Eqn. (3.19), that the zero-crossing variance increases as the square of the noise amplitudes, A_ν,

$$\sigma_l^2 \propto \frac{A_\nu^2}{A_l^2}. \qquad (3.31)$$

Thirdly, high-pass filtering of white noise is equivalent to differentiating it, by which higher frequencies get more power. Hence, the high-pass synaptic adaptation filtering further degrades the high frequency articulatory cues in presence of severe noise.

It is very common in HMM system to include delta coefficients in ASR parameterization. However, the calculation of delta features with the ZCPA do not provide encouraging results as in MFCC which may be due to the fact that the length of the time-window for the histogram construction is channel dependent, that is it varies inversely with the characteristic frequency of the channel. Thus appropriate delta features cannot be obtained with the derivative window lengths varying from 30-80 ms duration [7]. Therefore, delta features have not been used with ZCPA in this research. This is one of the motivations of using the synaptic adaptation as dynamic feature in the ZCPA.

We implemented a simplified model with fewer number of filters which may result in greater frequency overlap of adjacent frequency channels. This may introduce a histogram bias in the extracted features in a isolated word recognition task, resulting in a lower frequency resolution. For example, in 3.3 the recognition rates in Volvo noise for 15 dB SNR is observed to be the same for 10 dB SNR. Hence, a higher processing time with fewer number of filters is obtained at the cost of lower resolution with the isolated digits recognition. It is also useful to consider the computational cost of the temporal synaptic adaptation process to see whether it balances the gain in the recognition rates achieved. The synaptic adaptation filtering performed 4 multiplications, 4 additions and 1 division in each filter subband. At 30 dB SNR, a 9.1% improvement in recognition rate is obtained with an approximately 10% increase in processing time.

It is observed form Table 3.3 that many recognition rates are exactly equal for different SNRs and noise types. This is due to use of isolated digit recognition with male speakers only with a small number of models (eleven), and a very low vocabulary (eleven words) for the eleven digits with definite word boundaries. This produces a bias on misrecognition process. Usually, the misrecognised word is the same utterance across all speakers, such that the misrecognised words are always a fraction of 11, that is, 10/11, 9/11, 8/11, etc. for one, two and three words misrecognised, respectively. The isolated words recognition has been used primarily to measure the relative performances of two different algorithms, and the difference in results have been tested for statistical significance.

3.4 Summary and conclusions

In this chapter, a simplified scheme for ASR to include auditory adaptation in a zero-crossing algorithm utilizing temporal processing of speech is implemented. A first order high-pass IIR filter is used to replicate auditory adaptation with an synaptic adaptation time constant of 250 ms in a ZCPA auditory model. It is demonstrated that the rapid synaptic adaptation could be implemented in temporal processing, not otherwise possible in spectral domain processing. The ZCPA with temporal synaptic adaptation was further tested in a CV discrimination task using the UCLA-SPAPL CV speech corpus. It is observed that the largest percentage increase in the zero-crossing rate were recorded for the vowel /ee/ in all consonant types, followed by the vowel /oo/ in fricatives and affricates. A qualitative measure of vowel clustering with synaptic adaptation is obtained by two-dimensional grouped scatter plots of utterance matrix. A quantitative measure of vowel class separability is obtained from within class and between class scatter matrices using a linear discriminant analysis. It is observed that ZCPA_ADP showed greater vowel class separability than the base ZCPA for three vowels in syllable-initial, voiced plosive context.

The benefit with temporal synaptic adaptation was found to be marginal, with a small improvement in Gaussian and factory noise. On HMM recognition, no improvement is observed for the ZCPA with synaptic adaptation in clean condition. In presence of noise, improvement is observed in stationary white Gaussian noise at high SNRs, but worse performance for nearly all other noise sources. In particular, the effect with non-stationary factory noise was observed to be reciprocal to the Gaussian white noise, that is, degradation at high SNRs with improvement at low SNRs. In babble and Volvo noise, no improvement or a degradation is observed.

A classical frame based method embedded in a hidden Markov model (HMM) technique was used to evaluate temporal features/effects represented by the ZCPA model outputs. The classical frame-based framework (typically 25 ms) is based on the assumption that the features are stationary (or quasi stationary) within the frame window (duration). There may be some model-feature mismatch between the frame-based approach and auditory events like rapid adaptation, and also due to the wider window lengths (30 ms to 120 ms) employed in auditory processing with some loss of assumption of stationarity. As a result, a frame-based approach for HMMs in auditory processing has always been a compromise. Nevertheless, frame-based HMM technique for ASR in auditory models have been shown to provide improved performances as demonstrated by several researchers [16], [24], [7], [27].

It is expected that exponential increase in the processing speed and continued research in the future may alleviate some of the computational complexities associated with computational auditory models. Moreover, the proposed algorithm has been designed for hardware implementation, which may further reduce processing time such that real time applications may be possible.

Chapter 4

Speech recognition with two-tone suppression

In this chapter, the auditory phenomenon of two-tone suppression is investigated for performances in automatic speech recognition. Two-tone suppression is the reduction in amplitude of a weaker intensity tone (suppressee or probe tone) due to the presence of a stronger intensity second tone (suppressor tone) at a frequency higher or lower than the suppressed tone [52], [103]. It is a nonlinear property of the cochlea in the auditory system which plays an important role in the process of synchronization of the auditory nerves with the formant frequencies at different stimulus intensities [103]. In particular, suppression of the neural rate response to the second and the third formants in speech by an intense first formant may be a factor in speech perception [104]. Nonlinearity introduced by a companding (compression followed by expansion) strategy may be used to implement two-tone suppression and thereby enhance spectral contrasts in speech processing, particularly in hearing aid devices [105]. In this research, the companding strategy is applied to reproduce two-tone suppression in a temporal auditory model using a zero-crossing algorithm and evaluated for speech recognition performances in various noise types.

4.1 Two-tone suppression in the auditory system

Neural two-tone suppression originates in the mechanical phenomena at the BM due to the interaction between the outer hair cells and the BM [103]. Lyon and Mead [11] has proposed that although the main source of two-tone suppression is cochlear nonlinearity, the nature of that nonlinearity is not "saturation". Rather, it is more like an automatic gain control in which the mechanical gain gradually reduces as a function of the response level over a very wide dynamic range of input power levels [11]. The gain control or multichannel compression by itself improves audibility. For a probe tone located at a particular CF, and a suppressor tone with frequencies both higher and lower than the CF, the magnitude of suppression increases monotonically with suppressor intensity [103], [52]. Moreover, the rate of growth of suppression magnitude with suppressor intensity is higher for suppressors in the region below CF than for those in the region above CF [103]. Suppression has also been observed for a probe embedded in a band of noise [59].

Although, the gain control or multichannel compression improves audibility, the asymmetric amplification due to compression may degrade spectral contrast. The degradation of the spectral contrast due to compression may be prevented by introducing two tone suppression, which may improve spectral contrast and thus increase audibility in cochlear-implant processors and hearing aids through low-power analog VLSI implementations [105]. The spectral enhancement property of the two-tone suppression indicate that it may also be usefully utilized for ASR performance enhancement.

The cochlear model proposed by Kates [106] implemented a traveling wave amplification with a cascade of adaptive-Q filter sections to reproduce two-tone suppression, as observed in the biological cochlea. But such a system is not only complex, but causes interactions between the feedback and the feedforward parameters in a traveling wave architecture.

An alternate methodology, the companding strategy, was proposed by Turicchia *et al.* [105], which performed simultaneous multichannel syllabic compression and spectral contrast enhancement via two-tone suppression. Thus, the companding strategy may be effectively used in an ASR front-end for improving recognition performance. In our implementation, the companding architecture was implemented in a ZCPA auditory model.

4.2   The companding architecture

In comparison to spectral domain implementations, temporal companding implemented in the time domain has the following advantages. Firstly, temporal characteristics such as synchrony and phase lock properties may be captured in time-domain processing, as processed in the human auditory system. Secondly, it involves simple computations and transformation instead, of complex transformations from the time domain to frequency domain, and vice versa. The temporal companding system utilizes nonlinear interactions between a compression block and a expansion block to produce two-tone suppression effects [105]. It utilizes noncoupled filterbanks and compression-expansion blocks to reproduce the two-tone suppression and compressibility effects of the biological cochlea by using a set of independent and programmable parameters. The concept of companding has been applied in other fields of speech processing, particularly for improving the signal-to-noise ratio in cases where the dominant noise occurs after the compression block [107]. It has also been applied in novel analog filtering circuits [108], [109]. For example, companding using analog filters has been used in stereo audio systems in which the compressor provides variable pre-emphasis of high frequency signal components and the expander provides complementary variable deemphasis.

The companding architecture is shown in Fig. 4.1, the detailed implementation of which is given in [105]. Every channel has a pre-filter, F, a compression block, a post-filter, G, and an expansion block as shown in Fig. 4.1. The two-tone suppression effect is produced due to the presence of the second filterbank between the compression and the expansion blocks, and the nonlinear interaction between signals in the first filterbank, the compressor, and the second filterbank. The pre-filter and the post-filter have the same perceptual frequency scale, but the first filter is broadly tuned while the second filterbank is sharper tuned. Both the compression block and the expansion block consist of a feed-forward path made up of an envelope detector (ED), a nonlinear block and a multiplier. The envelope detector consists of a half-wave rectifier followed by a low-pass filter. The dynamics of the compression and the expansion is determined by the time constant of the envelope detector (τ_{ED}), which is scaled with the resonant frequency, f_i, for the channel i such that $\tau_{ED}=w\tau_i$, where $w=10$ and $\tau_i = 1/(2\pi f_i)$.

The amount of nonlinear compression and expansion are determined by the parameters n_1 and n_2, respectively. It is seen from Fig. 4.1 that the exponent n_1 in the compression block determines the compression when $n_1 < 1$. At the same time, if n_2 in the expansion block is chosen such that $0 < n_2 < 1$, the overall system be-

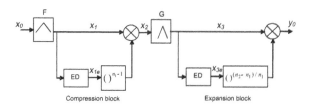

FIGURE 4.1: A single companding channel in the filterbank for implementing the two-tone suppression effect. Compression takes place for $n_1 < 1$, and expansion for $n_2 > n_1$ [105].

haves as a compression only system without any expansion effects. The amount of
compression for this case is given by the value of n_2. Expansion due to the presence
of the expansion block takes place only when the ratio n_2/n_1 is greater than 1 (that
is, when $n_2 > n_1$) and $n_2 ¿ 1$. Representative values for n_1 and n_2 for a companding
system are 0.5 and 1.2, respectively.

Let the input, x_0, in Fig. 4.1 be represented by

$$x_0 = a_1 \sin(\omega_1 t) + a_2 \sin(\omega_2 t + \varphi_0) \tag{4.1}$$

where ω_1 and ω_2 are the frequencies of the two tones in x_0. If the gain and phase of
the pre-filter, F, at the frequencies ω_1 and ω_2 are given as

$$
\begin{aligned}
f_1 &= |F(j\omega_1)|, \quad f_2 = |F(j\omega_2)| \\
\varphi_1 &= \angle(F(j\omega_1)), \quad \varphi_2 = \angle(F(j\omega_2))
\end{aligned}
\tag{4.2}
$$

and for the post-filter, G, at the frequencies ω_1 and ω_2 are given as

$$
\begin{aligned}
g_1 &= |G(j\omega_1)|, \quad g_2 = |G(j\omega_2)| \\
\varphi_1 &= \angle(G(j\omega_1)), \quad \varphi_2 = \angle(G(j\omega_2))
\end{aligned}
\tag{4.3}
$$

then the output, y_0, is given in [105] as

$$
\begin{aligned}
y_0 &= \left[a_1^{n_2/n_1} (a_1 + f_2 a_2)^{n_2(n_1-1)/n_1} \right] \sin(\omega_1 t + \varphi_1 + \vartheta_1) \\
&= \left[a_1 \left(\frac{a_1 + f_2 a_2}{a_1} \right)^{(n_1-1)/n_1} \right]^{n_2} \sin(\omega_1 t + \varphi_1 + \vartheta_1).
\end{aligned}
\tag{4.4}
$$

The derivation of Eqn. (4.4) is given in Appendix B. It is seen from Eqn. (4.4) that
the presence of a second tone (suppressor) with amplitude a_2 suppresses the tone
with amplitude a_1. If only a single tone is present such that $a_2 = 0$, then for $n_2 = 1$,
the output has the amplitude of a_1 only.

4.3 Implementation of temporal companding using zero-crossing algorithm

In this section, the biologically inspired companding strategy, which performs simultaneous multichannel syllabic compression and spectral contrast enhancement via
two-tone suppression, is implemented in a ZCPA auditory model used as an ASR

front-end.

One significant difference between our implementation and the implementation by Turicchia *et al.* in [105] is that in [105], it is implemented in the spectral domain, whereas we implemented the companding in the time domain using a zero-crossing algorithm, which we term as temporal companding. The advantages of time domain processing are stated in Sec. 1.1. Some enhancements to the original companding model in [105] are also proposed by us in the temporal companding. The suppression model proposed in [105] utilized cascaded second order biquad filter sections with 50 channels, which were computationally expensive. In the ZCPA, the temporal companding was implemented using simple FIR bandpass filters with 16 channels for computational efficiency and hardware implementation, following the implementation procedure as stated in Sec. 3.1.2 for adaptation. In the companding model in [105], a final summation across all the channels were taken to obtain a summary output. This might cause interference due to phase differences across channels because of adjacent channel interferences in the filterbank outputs. The companding scheme alleviated this by applying the first order low-pass filter twice for obtaining zero phase, once in the forward time direction and once in the reverse time direction. In the ZCPA, instead of taking a final summation at the output across all channels, the usual ensemble interval histogram is constructed from the zero-crossing intervals for all the channels. Since the phase information is not considered in a frequency histogram construction for the ZCPA [7], this eliminated the additional filtering required to minimize the phase effects. The proposed temporal companding also extends the capabilities of the ZCPA auditory model with the perceptual feature of two-tone suppression.

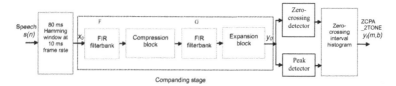

FIGURE 4.2: The ZCPA with two-tone suppression as an ASR front-end.

Fig. 4.2 shows the ZCPA model added with the two-tone suppression using the temporal companding method. In the companding stage, the 16 FIR filters in each filterbank F and G were spaced between 10-3500 Hz at the ERB scale, with the first filterbank bandwidths 20% higher than the second filterbank bandwidths. In each channel, the compression block in the companding stage consisted of an envelope detector (ED) which consisted of a half-wave rectifier followed by a first order low-pass filter. The filter time constant was made inversely proportional to the CF of the respective channel filter ($\tau_i = 1/(2\pi f_i)$). The half-wave rectifier used was given in [9] which combines the rectification process with the compressive nonlinear saturation effects, as seen in the biological cochlea,

$$
\begin{aligned}
y &= 1 + A \tan^{-1} Bx, & x &> 0 \\
&= e^{ABx}, & x &\leq 0
\end{aligned}
\tag{4.5}
$$

where x and y are the input and output, respectively, and A=10 and B=65 are constants equivalent to gain. It is seen that this function is exponential for negative inputs, linear for small input values and compressive for larger signals.

The half-wave rectifier, though removing the negative values from the speech signal, does not affect the zero-crossing feature extraction, due to the presence of the forward path after the FIR filtering and due to the presence of the multiplier in Fig. 4.1. The values of compression parameter n_1=0.5 and the expansion parameter n_2=1.1 were used. For these parameters, the extent of compression applied is (n_1-1)= -0.5, as observed in the compression block in Fig. 4.1. At the ZCPA output, an interval histogram, y_t, was constructed from the zero-crossing intervals from the output of the temporal companding block, y_0, which were collected in 26 frequency bins spaced at the ERB scale between 10-4000 Hz,

$$
y_t(m, b) = Z\{y_0\} = \sum_{i=1}^{C} \sum_{k=1}^{D_i-1} \log(P_{ik} + 1)\delta_{bj_k}, \qquad 1 \leq b < R
\tag{4.6}
$$

where $Z\{.\}$ is an operator for transformation by the zero-crossing algorithm, C the number of bandpass cochlear filters, D_i is the number of zero-crossings in the m-th frame and the i-th filter. δ_{bj} is the Kronecker delta as defined in Eqn. (3.17) and R is the number of histogram bins, weighted by peak values P_{ik} within a zero-crossing interval. Thirteen cepstra were collected from each frame at a frame rate of 10 ms.

Using Eqn. (3.19) and Eqn. (4.4), a relationship between the variance, σ_l^2, of the time interval perturbation of the zero-crossings, and the compression and expansion parameters, n_1 and n_2, respectively, may be obtained. For a two-tone input with amplitudes a_1 and a_2 with the suppressor tone, a_2, fixed at a CF at frequency l, and the probe tone, a_1, at a frequency close to l, the variance of the zero-crossing perturbation may be expressed, using Eqn. (3.18) for the the zero-crossing variance, as

$$\sigma_l^2 = \frac{2A_\nu^2 B_l / W (N-1)^2}{\left[a_1 \left(\frac{a_1 + f_2 a_2}{a_1} \right)^{(n_1-1)/n_1} \right]^{2n_2} \pi^2 E\{D_l\}^2} \tag{4.7}$$

where A_ν is the noise amplitude, W is the bandwidth of the noise power spectrum, $E\{D\}$ is the expected value of the zero-crossing count, and N is the total number of samples from which D_l is computed. The filter bandwidth, B_l, may be approximated by the average bandwidth of the compression block filter and the expansion block filter. It is observed from Eqn. (4.7) that due to the presence of a second tone, a_2, in the denominator, the zero-crossing variance is reduced.

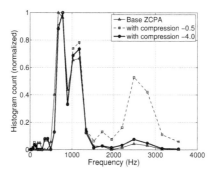

FIGURE 4.3: The first three formants of the vowel /a/ for the base ZCPA and ZCPA with temporal companding with two values of compression, -0.5 and -4.0.

Fig. 4.3 shows the synthesized short vowel /a/ for the base ZCPA and ZCPA with temporal companding for A=10 and B=65 with the two values of compression, -0.5 and -4.0 corresponding to n_1=1 and 4.5, respectively, maintaining the expansion same for both the cases. It is observed that with a compression of -0.5 as used in [105], the dominant first formant at 730 Hz is slightly suppressed compared to the base ZCPA, the second formant at 1090 Hz is enhanced, and so is the third formant at 2240 Hz to a much larger extent, which degraded the formant contrast with respect to the first and the second formants. The third formant is enhanced due to the fact that the higher gain provided by the A and B parameters in the half-wave rectification increased the peak values contributing to the histogram counts. The high frequencies are affected more than the lower frequencies due to the compression and expansion effects of the companding algorithm. We, therefore, increased the compression from -0.5 to -4.0, keeping n_1 and n_2 parameters the same as before (0.5 and 1.1 respectively). This was achieved by changing the exponent for the compression block in Fig. 4.1 from $(n_1\text{-}1)$ to $(n_1\text{-}4.5)$. This would have the effect of increasing the compression without affecting the expansion. It is observed that increasing the compression from -0.5 to -4.0, the second formant is slightly reduced while the third formant is significantly reduced, with an overall effect of improving the spectral contrast over the baseline ZCPA. Hence, a compression of -4.0 was used in all of our experiments.

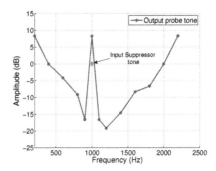

FIGURE 4.4: Two-tone suppressed output of a companding stage when excited with an input consisting of a pair of sinusoids (suppressor and a probe tone). A compression of -4.0 was used in the companding with A=10 and B=65.

Fig 4.4 shows the two-tone suppressed output of the temporal companding stage when excited with an input consisting of a pair of sinusoids (suppressor and a probe tone). The input suppressor tone was fixed at 1 kHz with a constant amplitude at 0 dB while the frequency of the input probe tone was varied from 100 Hz to 3 kHz keeping the amplitude constant at -20 dB. A compression of -4.0 was used in the companding. At the output of the companding stage, the probe tone, converted to dB. demonstrated suppression effects in the vicinity of the suppressor tone for frequencies both lower and higher than the probe tone frequency. The output probe tone has a gain effect due to the gain parameters A and B provided by the half-wave rectification given in Eqn. (4.5).

Since with two tone suppression, there is an aggregate reduction of firing rate due to the presence of a suppressor tone, the ZCPA output is expected to reflect this as a reduction in the histogram count in the vicinity of the frequency of the suppressor. Table 8.3 shows the data with the same setup as in Fig. 4.4 but the readings taken as the sum of counts for all the bins at the output of the ZCPA as the frequency of the probe tone is varied. The histogram shows a similar trend, that is, there is a reduction in the histogram counts as the probe tone approaches the stronger suppressor tone fixed at 1 kHz.

TABLE 4.1: The two-tone suppression effect on the ZCPA interval histogram count. It is observed that there is a reduction of histogram counts in the vicinity of the suppressor as the probe tone approaches the stronger suppressor tone fixed at 1 kHz.

Probe frequency f_2, Hz	ZCPA histogram count
500	17941
800	17511
1000	17618
1200	17545
1500	17770
1800	18745

Fig. 4.5 shows the time-frequency plots of the ZCPA interval histogram with and without the two-tone suppression effects in Figs. (a) and (b), respectively, for the utterance of the digit 'one' in clean conditions. The magnitude inhibition due to the suppression effects, which depends on the relative stimulus intensity within the utterance, is seen along the vertical z-axis. It is seen in Fig. 4.5 that the suppression is more prominent at lower frequencies (higher histogram bins) due to the higher intensities.

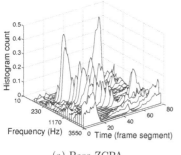 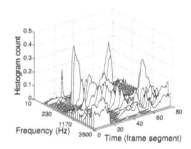

(a) Base ZCPA   (b) ZCPA with 2-tone suppression

FIGURE 4.5: The time-frequency plots of the ZCPA histogram for (a) ZCPA and (b) ZCPA with two-tone suppressed output for the digit utterance 'one' in clean condition. The magnitude inhibition may be observed along the z-axis (vertical) at lower frequencies (higher histogram bins).

4.4 A comparative analysis of CV discrimination with synaptic adaptation and two-tone suppression

In ASR applications, the classification of the vowels is more critical than the consonants [28]. In this section, the synaptic adaptation and the two-tone suppression are compared for their discrimination of vowel sounds using the UCLA-SPAPL CV corpus (Sec. 3.2). For comparisons, we define the frequency discrimination (FD) coefficient as

$$FD = \frac{\sum_{f>1340Hz} \text{ZC}(f)}{\sum_{f\leq 1340Hz} \text{ZC}(f)} \tag{4.8}$$

where ZC is the zero-crossing count as obtained from the interval histogram. A higher value of FD would indicate a high frequency enhancement or a low frequency suppression of a CV utterance, and vice versa. The high and low frequency discrimination was based on a frequency of 1340 Hz. The choice of this frequency was rather arbitrary and was partly based on considerations stated earlier, that it should be higher than the first formant, which is the dominant formant of the common vowels with maximum energy compared to higher formants. Another consideration for not using a higher frequency than the chosen one is that at higher frequencies, other high frequency artefacts of speech such as consonants (voiced and unvoiced plosives and fricatives) play a more important role, and may compromise the experimental results. The comparison of the FD values for the two algorithms of synaptic adaptation and the two-tone suppression in clean conditions is given in Table 4.2.

It is observed that both synaptic adaptation and two-tone suppression has higher FD values than the base ZCPA for all articulations. Moreover, in all cases, the vowel /ee/ has higher FD values than the vowels /a/ and /oo/. This is consistent with results given in Sec. 3.2.1, where it is shown that the percentage increase in the histogram weights above 1340 Hz with synaptic adaptation was higher for /ee/ than for /a/ and /oo/.

However, a significant observation made in Table 4.2 is that for the vowel /ee/, the increase in FD was higher with synaptic adaptation than with two-tone suppression (3.33 vs. 2.96 for /bee/), while for vowels /a/ and /oo/, the increase in FD is higher with two-tone suppression than with synaptic adaptation (0.23 vs. 0.40 for /ba/ and 0.51 vs. 0.78 for /boo/). These results indicate the frequency dependence of the perceptual features in speech processing, with the synaptic adaptation having

TABLE 4.2: Comparison of frequency discrimination coefficient, FD, of CV utterances between synaptic adaptation and two-tone suppression using the ZCPA.

	FD				FD		
	Base ZCPA	ZCPA _ADP	ZCPA _2TONE		Base ZCPA	ZCPA _ADP	ZCPA _2TONE
/ba/	0.22	0.23	0.40	/SHa/	0.58	0.65	0.95
/bee/	2.82	3.33	2.96	/SHee/	2.89	3.43	3.41
/boo/	0.46	0.51	0.78	/SHoo/	1.97	2.28	2.56
/pa/	0.28	0.30	0.46	/ja/	0.56	0.62	0.86
/pee/	2.31	2.79	2.53	/jee/	2.51	2.98	3.19
/poo/	0.45	0.50	0.74	/joo/	1.45	1.64	1.96
/ma/	0.33	0.36	0.63				
/mee/	1.92	2.32	2.10				
/moo/	0.43	0.47	0.73				

a greater effect on high frequency articulation, and the two-tone suppression having greater effect on low-frequency articulation. One exception to this rule is found for /jee/ where two-tone suppression performed better than synaptic adaptation. This may be due to the fact that this CV contains substantial frication in addition to the higher formants and the increase is mainly due to the reduction in the low frequency counts.

4.5 HMM word recognition with two-tone suppression

Table 4.3 shows the recognition results of the ZCPA with the two-tone suppression with speaker independent male isolated digits using the TIDIGITS corpus. The results with synaptic adaptation is also shown in the same table for comparison. There were 55 male speakers in the training set, each speaker with 2 utterances of the digits 1-9, 'oh' and 'zero' (total 1210 utterances in the training set).

A separate set of 31 male speakers in the test set each speaker with 2 utterances of the digits 1-9, 'oh' and 'zero' (total 682 utterances in the test set). The continuous Gaussian density HMM model parameters and the test parameters were the same as described in Sec. 3.3 for the case of synaptic adaptation. Test speech was corrupted with 4 different types of additive noise from the NOISEX 92 database: Gaussian white noise, factory noise, babble noise and Volvo noise. For the temporal companding, a compression of -4.0 and $n_2 = 1.1$ were used.

TABLE 4.3: Continuous density HMM recognition rates (%) of the ZCPA with two-tone suppression compared with the base ZCPA and ZCPA with synaptic adaptation for isolated digits (TIDIGITS corpus) with independent male speakers in four types of additive noise.

SNR (dB)	White			Factory		
	ZCPA	ZCPA _ADP	ZCPA _2TONE	ZCPA	ZCPA _ADP	ZCPA _2TONE
Clean	95.4	95.4	95.4			
40	90.9	95.4	95.4	95.4	90.9	95.4
30	81.8	90.9	54.5	86.3	81.8	95.4
15	77.3	72.7	50.0	81.8	77.3	90.9
10	68.2	59.1	45.4	68.8	72.7	86.3
5	50.0	31.8	40.9	50.0	68.8	72.7

SNR (dB)	Babble			Volvo		
	ZCPA	ZCPA _ADP	ZCPA _2TONE	ZCPA	ZCPA _ADP	ZCPA _2TONE
40	90.9	90.9	95.4	95.4	95.4	95.4
30	86.3	86.3	90.9	90.9	90.9	95.4
15	72.7	72.7	81.8	86.3	86.3	95.4
10	63.6	59.1	68.1	86.3	81.8	95.4
5	31.8	54.5	54.5	81.8	77.3	90.9

It is observed from Table 4.3 that the ASR performance with two-tone suppression follows the similar trend as with synaptic adaptation and the base ZCPA in [7]. That is, the best performance is obtained with Volvo noise, followed by factory noise. However, it is observed that with synaptic adaptation, white noise performs better than babble noise, whereas with two-tone suppression babble noise performs better than white noise. The reason for this may be explained using the analysis given in Sec. 3.3 where it is stated that in general, temporal synaptic adaptation performs better over the base ZCPA in stationary Gaussian white noise due to the zero crossing variance reduction effects for the higher frequency noise components, as observed from Eqn. (3.18) and Eqn. (3.19).

It is observed that the recognition results with two-tone suppression are improved over the base ZCPA and the ZCPA with synaptic adaptation under all noise types and over all SNRs, except for white Gaussian noise at moderate and low SNRs. A significant observation made from the Table 4.3 is that synaptic adaptation performs better than two-tone suppression in stationary white Gaussian noise and worse in non-stationary non-Gaussian noise (factory and babble), whereas an opposite effect is observed with the two-tone suppression, that is, two-tone suppression performs better in non-Gaussian real-world noise (factory, babble, Volvo) and worse in Gaus-

simulated in the speech parametrization by a pre-emphasis operation at the input. Although this is a simple scheme, the pre-emphasis increases the energy of the high frequency components but does not simulate the actual perceptual behavior across the entire audible frequency range. An alternative method would be to determine the coefficients for each equal loudness curve for the entire audible frequency range by some normalization procedure, and then use a lookup table to determine the coefficients for an applicable loudness level. The coefficients are multiplied with the mel log spectral magnitudes to adjust the spectrum to the human perceptual curves. A similar method of equal loudness normalization is used in multichannel dynamic compression applied in hearing aids to improve intelligibility, usually with a short time constants. The gain characteristics used for such dynamic compression are deduced from categorical loudness pre-emphasis scaling similar to the method used in the PLP processing [117].

A representative example of equal-loudness coefficients for several phons levels for 16 channels are tabulated in Table 5.1. The set of coefficients, denoted by $\{L_e\}$, were determined for a frequency range of 10-3500 Hz corresponding to the CFs of the filterbank implementation spaced at ERB scale. The coefficients were normalized to a base frequency of 10 Hz, which is taken as unity. For MFCC parametrization using a mel scale (instead of the ERB scale used in Table 5.1), a similar set of coefficients for mel frequencies, $\{L_m\}$, using 40 triangular filters (13 linear and 27 log), was determined which are given in Appendix C.1. These values are used in the experiments defined in this chapter.

The table is arranged as filterbank coefficients as rows and the equal phons levels as columns. A row of coefficients for a particular phons level may be determined by a lookup procedure. First the intensity level is determined at 1 kHz, which is by definition equal to the phons level. The coefficients corresponding to this phons level are applied as multiplicative gains to the existing filter coefficients. For ASR purposes, the table lookup procedure may be further simplified considering that the inputs are usually normalized between ± 1 to eliminate the effects of loudness. Therefore a single normalized curve corresponding to conversational speech, that is, for an intensity level of 60 dB at 1 kHz corresponding to a phons of 60, may be applied, which was followed in our experiments.

It is observed that the equal loudness coefficients thus obtained are functions of the centre frequencies of the filterbank, and therefore depend on the CFs and the number of filters used in the implementation. Therefore, we also determined a generalized analytic expression for the equal loudness coefficients for the 24 Bark frequencies independent of the filterbank implementation. Such an analytic expression for a set of coefficients, $\{L_b\}$, was obtained by a curve fitting procedure utilizing the unconstrained nonlinear optimization which finds the minimum estimate of a scalar function of several variables, starting at an initial value [118]. A fitted curve to the intensity-loudness curve for the threshold of quiet (corresponding to 0 phons) is determined from these estimates by a trial and error procedure. This expression as a function of the Bark number, B, may be expressed as

$$L_b(B) = [10A\exp(-\lambda B^{0.39}) + 0.027\tan(0.105B) + 7 \times 10^{-7}\exp(B^{0.9})] \quad (5.1)$$

where A=33.19, $\lambda = 2.15$ and B is Bark number (B=1,...,24). The perceptual critical band numbers B are related to the frequencies f using the transformation [57]

$$f = 650\sinh\left(\frac{B}{7}\right). \quad (5.2)$$

Eqn. (5.1) is a close approximation to the threshold of quiet, which may be translated to a desired phons level by an appropriate multiplying factor. However, in this experiment, we used the coefficient set $\{L_m\}$ corresponding to the 40 channel mel filtrebank given in Appendix C.1.

5.3 Asymmetric compression in ASR parametrization

5.3.1 Static logarithmic compression in the MFCC and the ZCPA

In the conventional MFCC parametrization, static compression is provided by a log operation on the FFT magnitudes after the mel filterbank binning. In the ZCPA it is provided by the logarithm of the peak value within a zero-crossing interval. In our experiment, two measurements were made to characterize and compare the static compression in the MFCC and the ZCPA parametrization: dependence of compression on frequency and dependence on wideband white noise.

Figs. 5.2 (a) and (b) show the I/O curves of the MFCC and the ZCPA, respectively, for three pure tone inputs of frequencies of 700 Hz, 1000 Hz and 3500 Hz and the digit 'one' in clean condition. The inputs (along x-axis) were normalized and scaled, ranging from 0 db SPL to 100 dB SPL, corresponding to 0.00002 Pascal to 2.0 Pascal, respectively. The outputs (along y-axis) were measured by summing the log magnitudes in the case of MFCC which is equivalent to loudness summation, and by summing the histogram counts in the case of ZCPA, for all channels for all

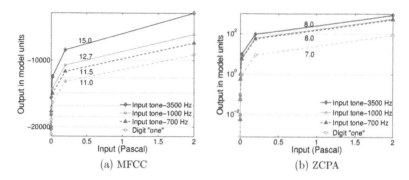

FIGURE 5.2: The I/O curves for three tones and the digit 'one' for an input range of 0.00002 Pascal to 2 Pascal (corresponding to 0 db SPL to 100 dB SPL), for the MFCC and the ZCPA showing the loss of compression at higher frequencies for MFCC, whereas ZCPA shows no such effects. The slopes are indicated with each curve as the angle with the x-axis in degrees for the input range of 0.2 - 2.0 Pascal.

frames. The output units, being different for the MFCC and the ZCPA, are shown in terms of the respective model units, which are the log spectral magnitudes for the MFCC and the histogram counts for the ZCPA. It is observed from the figures that for the MFCC, there is a loss of compression as the frequency is increased, indicated by a increase in the slope, whereas no such effects are observed for the ZCPA. The slopes in degrees (marked with each curve) are indicated as the angles each I/O curve makes with the x-axis for the input range of 0.2 and 2 Pascal.

Next the dependence of compression on white Gaussian noise was evaluated. The I/O curves for the MFCC and the ZCPA for the utterance of digit 'one' in clean condition and corrupted with white noise with SNR of 30 db, 10 dB and 0 dB are shown in Figs. 5.3 (a) and (b), respectively. Normalization in Fig. 5.3 was performed to limit the maximum values to -1 or +1 to demonstrate that normalization may be used to determine the difference in slopes from such figures. It is observed that as the noise level is increased, the compression in the MFCC is reduced, whereas for the ZCPA the compression is constant for the three values of SNR.

Motivated by the human auditory system, it is proposed that increasing compression in the MFCC in the high frequency regions may improve robustness of a speech recognition system. By asymmetric compression, degradation of the spectral contrast of the low frequency formants due to the added compression is avoided, resulting in improved audibility. This is consistent with human perceptual behavior that the compression is higher in the basal region of the cochlea which is more

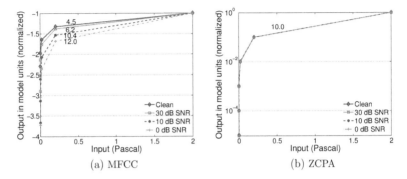

(a) MFCC   (b) ZCPA

FIGURE 5.3: The I/O curves for the utterance 'one' in white noise, showing loss of compression with increasing noise for MFCC, whereas ZCPA shows no such effects. The y-axis is normalized to contrast the slopes of each curve.

sensitive to higher frequencies [113]. Moreover, it may improve noise performance, particularly in white noise which contains substantial high frequency components and is also a major reason for the poor performance of ASR front-ends. However, care must be observed so that the asymmetric compression does not affect the frequency response of the tonal components.

5.3.2   The compression ratio and the compression angle

If $X(I)$ is the input and $Y(I)$ is the output as a function of the intensity I, then for a nonlinear I/O curve, the slope $k(X)$ may be expressed as $k(X) = \frac{dY(X)}{dX}$. We define the compression ratio (CR) as the output divided by the input, which is synonymous with the gain. In the cochlea, compression arises due to the nonlinearity in the gain characteristics due to a reduction in gain. In the nonlinear region, if the input intensity is incremented by ΔX and the output is incremented by $Y(X + \Delta X)$, the compression ratio, $g(X)$ may be defined as

$$g(X) = \frac{Y(X + \Delta X) - Y(X)}{\Delta X}. \tag{5.3}$$

In the limit $\Delta X \to 0$, it is seen that $g(X) = k(X)$, that is, CR is the same as the slope of the I/O curve at (X, Y). A general expression of a nonlinear system is

$$\frac{dY}{dX} = f(Y(X), X). \tag{5.4}$$

The characteristics of nonlinear elements may also include higher derivatives of the input signal and an analytic solution is usually difficult. However, in our case, an analytical expression was obtained by curve fitting using unconstrained nonlinear optimization using the I/O curve in Fig. 5.2 (a) corresponding to f=1000 Hz, as

$$Y = 100 + 0.2X - 8460X^{-0.12}. \tag{5.5}$$

The nonlinear Eqn. (5.5) may be made linear by first applying some transformation such as taking the logarithm of the input (along x-axis) in Fig. 5.2 (a). The slope of the linear log curve, that is, the tangent of the angle θ, denoted as the compression angle, with the x-axis may be interpreted as a representation of the compression associated with each curve. We first establish a reference compression curve as the curve for a 1000 Hz pure tone in clean condition, which is the condition when no external asymmetric compression is applied over the existing logarithmic mel compression. Such a reference curve is shown in Fig. 5.4, which corresponds to the curve in Fig. 5.2 (a) for f=1000 Hz with the input (x-axis) taken in log scale. The corresponding reference compression angle for the log I/O curve in Fig. 5.4 is determined as θ_r=36.97°.

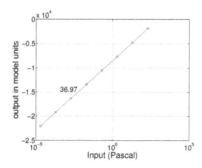

FIGURE 5.4: The linear log reference I/O curve for MFCC for a tone of frequency 1000 Hz in clean condition for an input range of 0.00002 Pascal to 2 Pascal in log scale. The reference slope is indicated in degrees as θ_r=36.97°.

5.3.3 The asymmetric compression coefficient

Assuming that the range of the compression angle is such that $0 < \theta \le 45°$, an asymmetric compression coefficient, C, may be defined, the value of which will depend on θ. The relationship between C and θ is generally nonlinear. C is multiplied to the mel log spectral magnitudes asymmetrically (at higher frequency channels) to obtain a desired compression. In the proposed scheme, the asymmetric compression was applied only to channel outputs with CFs higher than 1340 Hz. The choice of this frequency is based on consideration that it should be higher than the first formant, which is the dominant formant of the common vowels.

5.3.4   Application of asymmetric compression in the MFCC

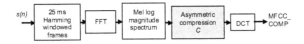

FIGURE 5.5: Schematic of the processing stages for the mel feature extraction using equal loudness normalization and asymmetric compression (MFCC_COMP).

Fig. 5.5 shows the preprocessing stages and the implementation of the proposed frequency dependent asymmetric compression method in the MFCC processing. Let $s(n)$ denote noisy speech such that

$$s(n) = m_s(n) + d(n) \tag{5.6}$$

where $m_s(n)$ is the clean speech corrupted by the additive noise $d(n)$. If $w(n)$ is a window frame of 25 ms duration, the value of $s(n)$ at the frame index, m, is given by

$$s(n; m) = s(n)w(m - n). \tag{5.7}$$

The k-point discrete Fourier transform (DFT) of $s(n; m)$ for each frame index m was computed as

$$S[k] = \sum_{n=0}^{N-1} s[n] e^{-\frac{j2\pi kn}{N}} \tag{5.8}$$

where N is the number of samples in a windowed frame. The mel magnitude spectrum was extracted as

$$S[l] = \sum_{k=0}^{N/2} S[k] M_l[k], \qquad l = 0, \ldots, L - 1 \tag{5.9}$$

where L is the total number of triangular mel filters and $\{M_l(k)\}_{k=0}^{N/2}$ are the filter coefficients. For our case, 40 mel triangular filters (13 linear and 27 log) were used. The set of coefficients $\{L_m\}$ for equal loudness normalization using the mel scale was obtained from a table similar to Table 5.1 for 40 channel mel frequencies, which is given in Appendix C.1. The log spectral magnitudes was multiplied by the coefficients $\{L_m\}$ corresponding to phons of 60 corresponding to normal conversational speech (column 6 of the Table 5.1 to obtain the log spectral magnitudes with equal loudness normalization, $\{S_e\}$, as

$$S_e[l] = \log S[l] \times L_m[l], \qquad l = 0, \ldots, L - 1. \tag{5.10}$$

$\{S_e\}$ was divided into a low frequency ($<$1340 Hz) segment, $\{S_{el}\}$, and a high frequency ($\geq$ 1340 Hz) segment, $\{S_{eh}\}$, such that

$$
\begin{aligned}
S_{el}[l] &= S_e[l] \qquad \text{for } l < l_0 \\
S_{eh}[l] &= S_e[l] \qquad \text{for } l \geq l_0
\end{aligned}
\tag{5.11}
$$

where l_0 is the channel with a characteristic frequency of 1340 Hz. In the next step, $\{S_{eh}\}$ was multiplied by C to implement the asymmetric compression. Eqn. (5.10) is the mathematical foundation of the proposed asymmetric compression which is used in the implementation of the computational algorithm.

5.3.5 Derivation of the compression angle

Since the relationship between C and θ is nonlinear, therefore, instead of an analytic solution the relationship is determined experimentally. That is, we have used several values of C, multiplied it with the mel log spectrum, and determined the corresponding linear log I/O curves which have been shown in Fig. 5.6. The compression angle was found from the slopes of each linear log I/O curve and has been tabulated in Table 5.2 for each value of C.

The range of C in relation to θ may be expressed as

$$0 < C \leq 1, \qquad \text{for } \quad 0 < \theta \leq \theta_r \tag{5.12}$$

where $C=1$ corresponds to the reference compression angle, θ_r, when no asymmetric compression is applied, and progressively decreases as the compression increases (that is, as θ decreases). Since C is multiplicative, a condition when $\theta=0$ would result in $C=0$, and must be avoided. Therefore, the lower range of C should extend

TABLE 5.2: Derivation of the compression angle and % compression from the compression coefficient, C.

Compression coefficient, C	Compression angle, θ	% compression
1.000 (Ref.)	36.97	0
0.698	33.23	10.12
0.443	30.58	17.28
0.216	26.31	28.83
0.031	23.14	37.41

from some very small value near zero, say 0.001, for maximum compression.

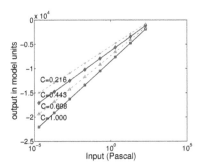

FIGURE 5.6: The linear log I/O curves corresponding to several C values. The linear curve corresponding to C=1.00 is taken as the reference curve.

The linear log I/O curves corresponding to several C values are shown in Fig. 5.6. It is found that for C=0.443 corresponding to θ=30.58°, which is an 17.28% increase in compression from the reference level, a good noise performance is achieved without too much degrading of the tonal components. Hence, the log mel spectrum after the frequency dependent asymmetric compression, $\{S_c\}$, is expressed as

$$S_c[l] = S_e[l] \times C, \qquad l = 0, \ldots, L - 1 \qquad (5.13)$$

where

$$C = \begin{cases} 1 & \text{for} \quad l < l_0 \\ 0.443 & \text{for} \quad l \geq l_0. \end{cases}$$

A DCT was performed on $\{S_c\}$ to extract 13 cepstra, including the zero-th coefficient, per frame at 10 ms frame rate.

Fig. 5.7 shows the I/O curves in clean condition without and with the application of equal loudness normalization and the asymmetric compression in Figs. (a) and (b), respectively, for three tones of frequencies 3500 Hz, 1000 Hz and 700 Hz. Comparing the two figures, it is observed in (b) that with the application of the asymmetric compression strategy (C=0.443), the slopes are decreased for all frequencies.

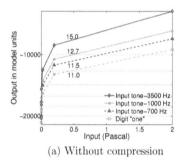
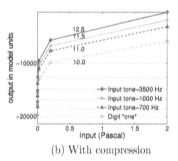

(a) Without compression   (b) With compression

FIGURE 5.7: I/O curves for the MFCC for three tones of frequencies 3500 Hz, 1000 Hz, 700 Hz and the digit 'one' (a) before and (b) after the application of equal loudness normalization and the asymmetric compression. The slopes are indicated as angles in degrees with each curve.

Similar reductions in slopes are observed in noise conditions with the asymmetric compression applied. These are shown in Figs. 5.8 (a) and (b) which show the I/O curves without and with the application of the the asymmetric compression (C=0.443), respectively, in clean condition and in white Gaussian noise.

Figs. 5.9 (a) and (b) show the spectrograms of the 32nd frame for MFCC for the utterance 'one' in 10 db SNR white noise without and with the application of equal loudness coefficients and asymmetric compression, respectively. The duration of the frame is 25 ms. That is, the processing started at 32nd frame and ended after that frame. A single frame was chosen for greater time-frequency resolution within this frame and comparing for the two cases of with and without asymmetric compression. It is assumed that the region of the utterance contain both low and high frequency tonal components. It is observed in (b) that in noise conditions, the low frequency segments are enhanced by the application of increased compression corresponding to C=0.443.

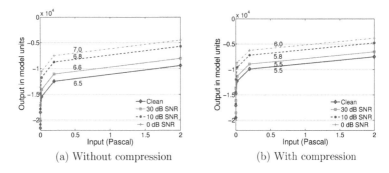

(a) Without compression   (b) With compression

FIGURE 5.8: I/O curves for MFCC for the utterance 'one' in white noise (a) without
and (b) with the application of the equal loudness normalization and the asymmetric
compression.

TABLE 5.3: Comparison of recognition rates (%) between the base MFCC_0 and MFCC_0
with equal loudness normalization and asymmetric compression (MFCC_COMP), using
male connected digits (TIDIGITS corpus) in clean condition and four types of additive
noise.

SNR (dB)	White		Factory		Babble		Volvo	
	MFCC_0	MFCC_COMP	MFCC_0	MFCC_COMP	MFCC_0	MFCC_COMP	MFCC_0	MFCC_COMP
Clean	94.21	98.28						
40	92.07	97.06	94.21	98.26	94.15	98.19	94.21	98.26
30	74.63	90.38	93.87	98.10	93.64	97.82	94.09	98.26
20	58.30	70.80	72.50	90.34	64.45	87.42	93.82	98.23
10	29.35	45.37	36.64	74.12	32.61	56.62	93.22	97.46

5.4 HMM word recognition with asymmetric compression in the MFCC

Speaker independent connected digits from the TIDIGITS speech corpus were used
for recognition experiments with the equal loudness coefficients and the asymmetric
compression in the MFCC. There were 55 male speakers in the training set and a
separate set of 31 male speakers in the test set, each with 77 connected digit utterances of the digits 1-9, 'oh' and 'zero'. The total utterances in the training set was
approximately 15670 and in the test set was approximately 7385 utterances. Continuous density HMM with 15 states per digit, 5 mixture components per state with
diagonal covariances were used to define each model. A 3-state silence model was
inserted at the beginning of each utterance. The Baum-Welch re-estimation using a
flat-start scheme and 15 estimation iterations per state was used for training under
clean condition. Test speech was corrupted with Gaussian white noise, factory noise,

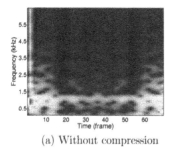

(a) Without compression

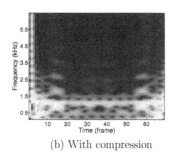

(b) With compression

FIGURE 5.9: Spectrograms of the 32nd frame for MFCC for the utterance 'one' in 10 db SNR white noise (a) without and (b) with the application of equal loudness normalization and asymmetric compression. In Fig. (b), it is seen that the low frequency segments are enhanced by the compression technique. (In gray scale, lighter shade indicates higher intensity and darker shade implies lower intensity).

babble noise and Volvo noise from the NOISEX 92 database. A left-to-right Viterbi recognizer was used for evaluating correct utterances. The word recognition rate was based on the percentage of accuracy, which takes into account the substitution, the deletion and the insertion errors.

Table 5.3 compares the recognition results between the baseline MFCC with the zero-th cepstral coefficient (MFCC_0) and MFCC_0 with equal loudness coefficients and asymmetric compressions (MFCC_COMP). The improvements in MFCC_COMP are observed in clean condition and in all noise types. For example, with a 17.28% increase in compression from the reference value at 10 dB SNR, the maximum improvement was observed in factory noise (37.48%), followed by babble noise (24.01%), white Gaussian noise (16.02%) and Volvo noise (4.24%). Thus, the improvements are more pronounced in presence of real-world, non-stationary noise types than in stationary noise types. This may be due to the increased low frequency formant contrasts, particularly in presence of low-frequency real-world noise conditions. It has been observed that the main contribution to the improvements is due to the asymmetric compression, rather than due to the application of the equal loudness coefficients. It is expected that the proposed method of asymmetric compression may yield further benefits to the state of the art parameterizations utilizing MFCC with CMN and dynamic (delta-delta) features.

In speech recognition tests, it is usually important to perform a statistical significance test to determine whether the differences in error-rates between two algorithms tested on the same data set are statistically significant. The McNemar's test [102] can be used for this purpose. The recognition results in Table 5.3 were further tested using the McNemar's test for the 682 test utterances and it was found that the differences in recognition results due to the two algorithms were statistically significant.

5.5   Summary and conclusions

Motivated by the human auditory system, we have characterized the logarithmic compression provided in the MFCC parameterizations by analyzing the psychophysical I/O perception curves. Particularly, we have investigated the frequency dependence of compression as a function of intensity levels. One important observation is that in clean condition, MFCC demonstrates a loss of compression at higher frequencies, particularly at high intensity levels, whereas no such effects are observed with the ZCPA. Same effect is observed in high white noise conditions. The insensitivity of the ZCPA compression to frequency and noise may be due to the fact that the interval histogram is less affected by frequency and noise parameters than the spectral magnitudes. Two frequency dependent compression strategies are presented: the application of the equal loudness coefficients and a frequency dependent asymmetric compression technique. In the asymmetric compression, the extent of compression applied determines the slope of the linear I/O curve which is obtained by taking the logarithm of the scaled input of the nonlinear I/O curve. The asymmetric compression has the advantage of increasing compression for improved audibility without degradation of spectral contrasts, and may be used as an alternative strategy to the companding scheme without the complex computations involved with it. The improvement results primarily due to the enhancement of the spectral contrast of the first formant compared to the higher frequency formants. On HMM recognition, performance improvements was observed in clean condition and in white, factory, babble and Volvo noise conditions, that is, in all noise conditions and at all SNRs. The improvements in the recognition performance in the MFCC are achieved without increasing the computational burden on the feature extraction.

Chapter 6

Feature extraction for ASR based on higher auditory processing

In this chapter, a method of feature extraction for ASR based on the auditory processing in the cochlear nucleus is presented. Up to the present time, considerable knowledge about the functionalities of the peripheral auditory system has been accumulated, primarily based on physiological measurements and psychoacoustic experiments on frequency selectivity, loudness perception, short-term synaptic adaptation, temporal masking, and two-tone suppression, which form the basis of many developed auditory models. Comparatively, little is known about the neural mechanisms of the central auditory processing stages [2]. It is believed that the initial process in the peripheral auditory system enhances some perceptual cues in the speech which assists the higher auditory system in the proper identification of the speech segments [28]. In response to the initial process, the auditory brainstem provides multiple representations of such information [37], [119].

In the conventional processing of speech by an auditory model, speech information is represented by the neural discharge of the auditory nerves, as shown in Fig. 6.1. When these models are employed as an ASR front-end, speech features are extracted from this discharge pattern with all perceptual features integrated into a single set of feature vectors, suitable for processing by a speech recognition system. In contrast, we propose a differential processing strategy as observed in the cochlear

FIGURE 6.1: Conventional feature extraction using an auditory model.

nucleus. Auditory processings are segregated at feature extraction level, similar to the processing strategy of the CN, from which features are extracted suitable for a recognition task.

In the auditory pathway, the auditory nerve (AN) fibres connected to the hair cells innervates a distinct region called the cochlear nucleus (CN), located in the central dorso-lateral side of the brainstem. The CN is the first nucleus in the central auditory system, as all auditory-nerve fibres connected to the IHCs terminates in the CN. After initial auditory processing, the CN makes multiple projections to the next higher auditory centers [120]. In contrast to the auditory nerves whose responses are monotype ("primary" response), there are up to 6 different responses observed in the CN [59]. In the CN, distinct regions are assigned to distinct tasks and its principle cells constitute separate, parallel processing pathways for encoding different properties of the auditory signal [119]. It is suspected that there are complex feature detectors in these higher stages of the auditory system, but little is known of their functional behavior [7].

Following this strategy, three different parallel processing algorithms, each one implementing a distinct subtask of the speech recognition system, were implemented. These are for the synchrony detection, the mean discharge rate processing, and the two-tone suppression across the frequency range, as observed in the anteroventral cochlear nucleus (AVCN), the posteroventral cochlear nucleus (PVCN) and the dorsal cochlear nucleus (DCN), respectively, of the CN. The synchrony primarily detects the low frequency formants in temporal phase lock to the input stimulus, and the mean rate is associated with the average rate of neural discharge and are more related to the broadband transients like onsets and other dynamic behaviours. The two-tone suppression performs somewhat like lateral inhibition [121], by suppressing the firing rates of the neighboring neurons, and thereby improving spectral contrast.

The mean rate path also implemented the perceptual property of rapid and short-term synaptic adaptation occurring in the synapses between the hair cells and the AN. An estimate of the mean rate of the auditory-nerve output may be obtained from the short-term average of the synapse outputs, which are continuous functions and proportional to the instantaneous rate of auditory-nerve discharge in a given frequency band [122].

The application of differential processing strategy in speech recognition is not new and has been applied in some forms in auditory models. Experiments from auditory physiology indicate that higher auditory processing make use of both the synchrony in the response to low frequency auditory-nerve fibers and the mean rate of response [9], [124]. Higher auditory processing is characterized by a strong syn-

116

chronization for low modulation frequencies, and a stochastic behavior for higher frequencies [123]. Seneff's joint synchrony/mean rate model [9] based on the generalized synchrony detector (GSD) provided two separate outputs for the synchrony and the mean rate, each of which represented spectral segments appropriate for distinct subtasks of a speech processing system. The model conformed with the experimental data obtained by Smith *et al.* [54] which demonstrated a constant ratio of onset to steady-state response.

Hunt *et al.* [15] modified Seneff's GSD model by adding adjacent-channel cross-correlation, which resulted in improved sensitivity to formants in noise conditions. This scheme allowed human frequency masking measurements to be replicated quantitatively. The modified GSD output demonstrated better response than the conventional mel filterbank when it was tested in a speech recognition task in clean and noisy conditions using the Euclidean distance spectrum measurement.

C. Kim *et al.* [122] combined the synchrony response with the mean rate response in a speech recognition experiment using an auditory model and the CMU SPHINX-III system. The recognition rate was observed to improve over the conventional MFCC with CMS, and over the mean rate output alone.

Xiang's *et al.* model [38] followed the concept of Seneff in the initial stages (stage I and II) and was more similar to CN processing than the implementations of Seneff and C. Kim. It utilized an averaged localized synchrony response (ALSR) [25] for synchrony detection. The firing rate was extracted by summing the probabilities of firing for all subband filters within a frame window. Additionally, it added the effects of lateral inhibition to each branch as weights applied to the ALSR and the firing rate outputs. The cepstrum of each branch was then computed. The model was tested for speaker identification in a multilayer perceptron classifier. It was shown that both ALSR and the firing rate responses performed better than the MFCC and the LPCC in noisy conditions. However, no HMM recognition performance was given by the authors.

6.1   The proposed CN processing strategy applied to ASR front-end

The method proposed by us has three distinctive features compared to the implementations mentioned above. Firstly, the proposed scheme was implemented using temporal processing of speech using a zero-crossing algorithm, which has several advantages over spectral processing, as stated in Chapter 1 (Sec. 1.1). The ZCPA auditory model was employed for this purpose, which utilized simple and fewer number of FIR filters for computational efficiency. The ZCPA, due to the temporal processing method and the dominant frequency principle, performs basically as a synchrony detector for the detection of the low frequency formants. In the base ZCPA, the auditory nerve firings are simulated by the positive-going zero-crossings of the input stimulus. The simulated discharge rate does not depend on auditory processing that take place in the synapses, such as the rapid and short-term synaptic adaptation. It is shown in Chapters 3 and 4 that adding synaptic adaptation and two tone suppression in the ZCPA may improve performances in noisy environment depending on the noise types. The mean discharge rate information is not emphasized in the ZCPA feature extraction, which may be one reason for the degradation of the ZCPA response in clean condition over the conventional feature extraction methods, such as the MFCC. The proposed scheme also extends the ZCPA features with mean rate (MR) information utilizing processes that take place in the synapses, such as the synaptic adaptation.

Secondly, instead of sequential processing of auditory features, a parallel processing strategy as observed in the CN was utilized. Two different filterbank and analysis windows for the synchrony path and the MR path suitable for the frequency range and the particular speech subtask being processed were utilized. The frame rates of the two filterbanks were synchronized at 10 ms.

Thirdly, unlike the GSD model of Seneff, synchrony detection was separated from the synaptic processing. This separation is partly motivated by the consideration that the synaptic outputs, which is more related to dynamic properties, do not provide a precise estimate of formant frequencies when displayed as a spectrogram [9], which primarily emphasizes the steady-state temporal variations, rather than the transients.

This separation is also a natural consequence of adapting the differential processing strategy of the CN. Another motivation is that half-wave rectification (HWR), which is used to simulate the motions of the BM into a sequence of spikes in the inner hair cells, introduces substantial higher order harmonics of the formant frequencies [9]. This degrades the formant detection, in addition to modifying the statistical parameters such as mean and variance which degrades ASR performance in both clean and noise conditions. Fig. 6.2 shows the HWR output (lower panel) showing the fundamental and two dominant harmonics to a 2500 Hz fundamental frequency tone input (upper panel) using Eqn. (4.5) for half-wave rectification, as given in [9].

An additional problem of half-wave rectification in the ZCPA is that half-wave

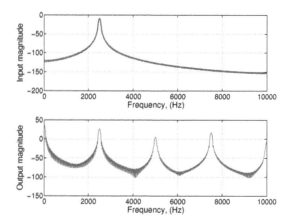

FIGURE 6.2: Half-wave rectification output (lower panel) using Eqn. (4.5) with compressive non-linearity showing the fundamental and two dominant harmonics to a 2500 Hz fundamental tone frequency input (upper panel).

rectification removes the negative portion of the signal, which removes the zero crossings, and hence, histogram construction is not possible. The mean is also raised from the zero value, but ZCPA feature extraction from level values higher than the zero level result in higher sensitivity in the estimated intervals and frequencies [7]. By this separation of the synchrony detection from the MR processing, the sensitivity of the synchrony output to spurious peaks which arise due to cochlear nonlinearities such as rectification is avoided. In ASR applications, the correct classification of the vowels is more critical than the classification for consonants [28]. By the proposed separation, the vowel formant detection remains unaffected by harmonic distortions, and phase locking to the harmonics of the dominant frequency is also avoided.

6.2 The central auditory system and the cochlear nucleus

Fig. 6.3 shows the afferent (ascending) and efferent (descending) pathways to and from the right ear peripheral auditory system to the auditory cortex. A similar pathway exists for the left ear. The auditory nerves attached to the organ of corti are connected to the next higher region called the cochlear nucleus. The CN is divided into distinct regions based on both physiological and anatomical criteria [59]. Specific regions of the cochlear nucleus are assigned to specific tasks related to pro-

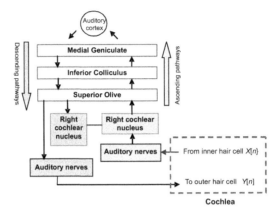

FIGURE 6.3: Ascending and descending pathways from the right ear to and from the higher auditory system. Similar pathways exist for the left ear.

cessing incoming speech stimulus. At least three major divisions of the nucleus can be distinguished on the basis of morphology, each of which contains a complete representation of the audible frequency range. Specific regions on the cochlear nucleus are assigned to specific tasks related to processing the incoming speech stimulus [59]. Some cells are excitory and some are inhibitory. Some pathways provide immunity to noise. Several parallel outputs radiate from the cochlear nucleus to the higher auditory regions leading to the auditory cortex.

The three main distinct regions of the cochlear nucleus are the anteroventral cochlear nucleus (AVCN) innervated by the ascending pathway, and the posteroventral cochlear nucleus (PVCN) and the dorsal cochlear nucleus (DCN), both innervated by descending pathways. Each region performs a specific task of the audible frequency range and plays a distinct and specific role in the perception of speech [59]. Twenty-two different types of neurons have been distinguished in the cochlear nucleus which can be grouped as primary-like, chopper, onset neurons and octopus cells, buildup and pauser, etc. [38]. Primary functional regions of the cochlear nucleus based on physiological criteria, the corresponding neuron types and the perceptual response patterns are shown in Fig. 6.4.

The AVCN neurons are mainly primary-like types which can maintain temporal-place code of AN fibres. The bushy cells in the AVCN have specialized electrical properties that allow them to transmit timing information from the auditory nerve to more central areas in the auditory system. These have been found to phase lock in a more precise manner than auditory nerve fibres [125].

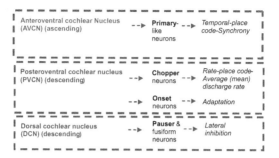

FIGURE 6.4: Processing regions of the cochlear nucleus based on physiological criteria, neuron types in each region and the response patterns.

The PVCN can maintain the rate-place codes in the auditory nerve fibres and is somewhat more complex and heterogenous in structure than the AVCN. Many more fibres terminate on each cell in the PVCN than on each cell in the AVCN. An AVCN neuron discharges each time an input discharge occurs, while for a PVCN neuron each discharge requires the accumulation of a number of input discharges [119].

Onset neurons (ON), also known as octopus cells in the PVCN, respond with great precision to signal onsets and broadband transients. These cells have a temporal resolution of neural coding of approximately 0.1 ms [126]. Another cell type found in the PVCN are the stellate cells, also known as the chopper cells, in reference to their ability to fire a regularly-spaced train of action potentials in tonal or noise stimulus. The firing rate depends on the strength of the auditory input more than on the frequency.

The DCN differs from the ventral portion of the CN as it not only projects to the central nucleus of the inferior collicus but also receives efferent innervation from auditory cortex, superior olivary complex and inferior collicus. Fusiform cells and the giant cells are the principal cells (also known as type IV cells) of the DCN. Current auditory models of the DCN employ the two-inhibitor model in which type IV cells receive excitation directly from the auditory nerve, and are inhibited by type II (vertical) cells and a wideband inhibitor (onset-c cells) [127], [128]. Thus it performs more complex auditory processing, rather than merely transforming information. The two-tone suppression curves are reminiscent of the consequences of lateral inhibition used in speech enhancement [105], [38].

Instead of a detailed modelling of the CN, we have attempted a broad functional representation of the CN based on the above properties and processing strategies for application in an automatic speech recognition task. Fig. 6.5 shows the processing (computational) models corresponding to the three distinct regions of the cochlear nucleus. The primarily-like AVCN neurons can maintain the temporal-place code of AN fibres and may be modeled as the synchrony detector output, which can maintain the phase lock property below approximately 1.34 kHz. PVCN, mainly composed of onset and chopper neurons, can maintain the rate-place code in the AN fibres and may be modeled as the average (mean) discharge rate output derived from a half-wave rectification and rapid and short-term synaptic adaptation. The DCN, consisting of buildup/pauser neurons, have inhibitory effects similar to lateral inhibition, and may be implemented by two-tone suppression using a temporal companding scheme [105].

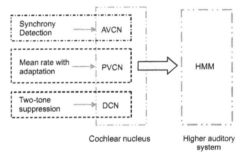

FIGURE 6.5: Computational models corresponding to the three distinct regions of the cochlear nucleus.

The feature extraction for ASR using the differential pre-processing is shown in Fig. 6.6. Three sets of 13 cepstra were generated from the parallel processing of each module from which a weighted composite 26-dimensional cepstrum feature vector per frame was constructed from the first 13, 7 and 6 cepstra from the synchrony, the two-tone suppression and the MR outputs, respectively. The weight of the MR processing was kept lower than the other two mainly due to its susceptibility to noise conditions.

6.3 The anteroventral cochlear nucleus (AVCN)

The AVCN processing is implemented by a ZCPA auditory model which primarily functions as a synchrony detector. The temporal processing of the ZCPA can maintain synchrony between the neural discharges and the input stimulus. At low CFs, the formants are resolved with high precision due to the dominant frequency principle. However, at high CF's, the frequency resolution is poor and the phase-locking

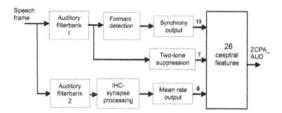

FIGURE 6.6: The weighted feature extraction (26-dimensional) using the differential pre-processingFtime-frequency. The numbers in the output of each module represent the cepstral features contributed by that module to the 26-dimensional composite feature vector.

of the discharges is greatly reduced due to the wider bandwidths. This is consistent with the human auditory system which has poor frequency resolution at higher frequencies. However, with the loss of frequency resolution, there is a corresponding increase in the time resolution at higher frequencies. The ZCPA is described in detail in Sec. 3.1.

The AVCN processing output is obtained after the filterbank utilizing a zero-crossing detector and a peak detector from which an interval histogram, $y_s(m, b)$, where m is the frame index and b is the interval histogram index, is constructed (Eqn. (3.16)),

$$y_s(m,b) = \sum_{i=1}^{C} \sum_{k=1}^{D_{s_i}-1} \log(P_{ik}+1)\delta_{bj_k}, \qquad 1 \leq b < R \qquad (6.1)$$

where D_{s_i} is the number of upward-going zero-crossings at the i-th filter in each frame.

The amplitude of the incoming speech was first normalized. The transduction of the mechanical vibration of the BM into frequency components was simulated by 16 finite impulse response (FIR) bandpass filters spaced at the equivalent rectangular bandwidth (ERB) scale between 10-3500 Hz. In each filter output, the inverse of zero-crossing intervals were collected in 26 frequency bins uniformly spaced between 10-4000 Hz at the ERB scale. The interval histogram was weighted by the logarithm of the peak values within a zero-crossing interval. One ZCPA frame was obtained every 10 ms. In the filter subbands, the zero-crossing intervals were collected over a derivative window length of $10/f_k$ for lower frequencies and $60/f_k$ for higher frequencies, where f_k was the filter center frequency. The histogram was normalized to reduce the effects of biasing. Thirteen cepstra were generated and retained from each speech frame.

6.4 The dorsal cochlear nucleus (DCN)

The DCN consists of the buildup/pauser neurons which have inhibitory effects similar to lateral inhibition, which are simulated by two-tone suppression using the temporal companding and the ZCPA, as implemented and explained in detail in Chapter 4. The effects of two-tone suppression for an utterance 'one' using temporal companding and the ZCPA may be observed in Fig. 4.5, in which the magnitude inhibition may be clearly seen along the z-axis (vertical).

When two tones of frequencies f_1 and f_2 with amplitudes a_1 and a_2, respectively, are present, the output of the temporal companding stage is given by Eqn. (4.4), which is repeated here as

$$y_0 = \left[a_1 \left(\frac{a_1 + f_2 a_2}{a_1} \right)^{(n_1-1)/n_1} \right]^{n_2} \sin(\omega_1 t + \varphi_1 + \vartheta_1)$$

where n_1 and n_2 are the nonlinear compression and expansion block parameters, respectively. This is input to the ZCPA from which an interval histogram, $y_t(m, b)$, as a function of frame index, m, and the histogram bin index, b, is constructed as

$$y_t(m, b) = \sum_{i=1}^{C} \sum_{k=1}^{D_{t_i}-1} \log(P_{ik} + 1)\delta_{bj_k} \qquad 1 \leq b < R \qquad (6.2)$$

where D_{t_i} is the number of upward-going zero-crossings, D_t, at the i-th filter in each frame and δ_{bj_k} is the Kronecker delta defined in Eqn. (3.17).The frame rate was synchronized with the synchrony detector. Thirteen cepstra were generated from $y_t(m, b)$ for each frame at a frame rate of 10 ms.

6.5 The posteroventral cochlear nucleus (PVCN)

While the synchrony output extracts detailed temporal information patterns about the frequencies present, the mean rate output of the auditory nerve fibres represents the gross temporal pattern due to dynamics of the synapse processing. The mean rate output may be obtained from the short-term average of the synapse outputs, which are proportional to the instantaneous rate of auditory-nerve firings in a given frequency band [7]. The nerve responses are not observed as spike trains, which would be the case for single neurons. Instead, the mean rate output represent the probability of firing as a function of time for an ensemble of similar fibres acting as a group. It determines the envelope amplitude, corresponding to the average discharge rate response and, therefore, may be obtained at the output of an envelope detector in a given frequency band per frame [9], [122] as shown in Fig. 6.7.

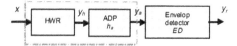

FIGURE 6.7: The IHC synapse model for the MR output, consisting of half-wave rectification (HWR), synaptic adaptation (ADP) and low-pass filter for envelope detection (ED).

The synaptic output provides a representation for locating transitions and discriminations of fricatives, closures, weak sonorants and vowels. Although the mean discharge response may contain frequency information, these are smoothed out by the subsequent stages. The MR path was separated from synchrony detection for reasons explained in Sec 6.1.

6.5.1   The half-wave rectification in the synapse processing

It was stated in Sec 2.3.2 that the inner hair cells act as half-wave rectifiers for the velocity of the motion of the fluid and spikes tend to be initiated on only half-cycle of the BM movement. For this reason, conventional auditory models utilize half-wave rectification in the synapse processing. In the MR path, which does not utilize a zero-crossing algorithm for feature extraction, the half-wave rectification used [9] is as given by Eqn. (4.5),

$$y_h = \begin{cases} 1 + A \tan^{-1} Bx, & x > 0 \\ e^{ABx}, & x \leq 0 \end{cases}$$

where x and y_h are the input and the output, respectively. The dynamic range compression as saturating nonlinearity, as well as half-wave rectification are performed by Eqn. (4.5).

6.5.2   The temporal synaptic adaptation - onset neurons

The onset neurons (ON), also known as octopus cells in the PVCN, are more responsive to changes than to steady inputs. Adaptation accentuates signal onsets by following a high initial firing rate. This corresponds to the activity of the octopus cells and the onset neurons (ON), which respond to signal onsets with higher discharge rates, gradually decaying to a steady-state value. Exact roles of ON in speech processing are unknown, but it has been hypothesized that by detecting signal onsets, the ON assist in encoding the fundamental frequency [120]. The simplified synaptic adaptation model for ASR using a high-pass IIR filter function as given in Eqn. (3.22) was utilized and is restated here for convenience as

$$H_a(z) = \frac{10\tau f_r(1 - z^{-1})}{(10\tau f_r + 0.05) + (10\tau f_r - 0.05)z^{-1}}$$

where τ=250 ms is the time constant, and f_r is the frame rate equal to 100 Hz. The adaptation model is explained in detail in Chapter 3, where it is shown that a high adaptation time constant improves the ASR performance in white noise at high SNRs. The synaptic adaptation output, $Y_a(z)$, is given by

$$Y_a(z) = Y_h(z)H_a(z). \tag{6.3}$$

where $Y_h(z)$ is the z-transform of the output of the half-wave rectification, $y_h(n)$.

6.5.3 The mean (average discharge) rate processing

We propose to utilize two different frequency bands and data window lengths in the synchrony and the MR paths to enhance the particular segments of the speech. The synchrony filterbank utilized the low frequency band of 10-3500 Hz with wider bandwidths and wider data windows suitable for detection of the vowel formants. The MR filterbank utilized a wider frequency band of 10-8500 Hz, and had a narrower bandwidth and smaller data windows for the detection of high frequency consonants, plosives and the fricatives. The narrower bandwidth also assists to limit the noise. The ZCPA synchrony detector utilized a frequency dependent derivative window Sec. 3.1.2 with a maximum width of 80 ms corresponding to the lowest CF channel. In the MR path, a constant derivative window length of 25 ms was used to increase the time resolution to capture the envelope profile due to synapse processing. The frame rate for the MR processing was synchronized with the synchrony detector at 10 ms.

For a single frame, each adaptation filter output, y_a, was filtered by a first order low-pass smoothing filter (ED) to detect the envelope of the signal using

$$y_r(n) = y_a(n) + 0.999y_r(n - 1). \tag{6.4}$$

The maximum value in $y_r(n)$ in each filter output, was retained as the frame data. This was accumulated for all frames m, to obtain $y_r(m, f)$

Each output in $y_r(m, f)$ may be considered as the short-term average discharge of all the fibres sensitive to a particular CF, and can be modeled probabilistically. Considering Y_r as a continuous random variable associated with each output with a probability density function $f(Y_r)$, then

$$\int_{\text{all } Y_r} f(Y_r)dY_r = 1 \tag{6.5}$$

and the expected value of Y_r is given as

$$E\{Y_r\} = \int_{\text{all } Y_r} Y_r f(Y_r)dY_r. \tag{6.6}$$

The variance is given by

$$\text{Var}(Y_r) = E\{Y_r^2\} - [E\{Y_r\}]^2. \tag{6.7}$$

Usually Y_r is multimodal with a finite number of Gaussian mixtures. If the number of mixtures is denoted by g, in proportions $\pi_1, \ldots, \pi_g$ then the p.d.f. of an observation Y_r can be expressed in finite mixture form as

$$f(Y_r; \phi) = \sum_{i=1}^{g} \pi_i f_i(Y_r; \theta) \tag{6.8}$$

where $f_i(Y_r; \theta)$ is the p.d.f. corresponding to g_i, θ denotes the vector of all unknown parameters associated with the parametric forms adopted for the g component densities, and ϕ denotes a vector of all unknown parameters associated with Y_r. The sum of the g mixtures in proportions $\pi_1, \ldots, \pi_g$ is given by

$$\sum_{i=1}^{g} \pi_i = 1, \qquad \pi_i \geq 0 \qquad (i = 1, \ldots, g) \tag{6.9}$$

6.6 The CN feature extraction model

Fig. 6.8 shows the detail components of the CN feature extraction model.

If $s(n)$ represent the input speech corrupted by additive noise, then the outputs shown in Fig. 6.8 are given by the following equations, where the parameters are defined in Sec. 3.1, Sec. 4.2 and Sec. 6.5.

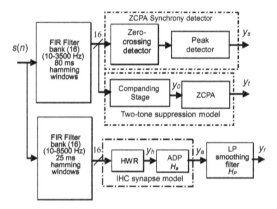

FIGURE 6.8: The detail components of the CN feature extraction model. HWR_ADP is the output of the IHC synapse model.

$$y_s(m,b) = \sum_{i=1}^{C} \sum_{k=1}^{D_{s_i}-1} \log(P_{ik}+1)\delta_{bj_k} \qquad 1 \le b < R$$

$$y_t(m,b) = \sum_{i=1}^{C} \sum_{k=1}^{D_{t_i}-1} \log(P_{ik}+1)\delta_{bj_k} \qquad 1 \le b < R$$

$$y_r(n) = y_a(n) + 0.999 y_r(n-1)$$

$$Y_a(z) = Y_h(z)H_a(z). \qquad (6.10)$$

$Y_a(z)$ is the z-transform of the synaptic adaptation filter output, $y_a(n)$, as shown in Fig. 6.8. Out of thirteen cepstra generated from each of the three sections, all 13 from the AVCN, 7 from the DCN and 6 from the PVCN were retained to form a 26 dimensional cepstral feature vector, C_m,

$$C_m = [C_{s_m}, C_{t_m}, C_{r_m}]^T, \qquad m = 1,\ldots,M \qquad (6.11)$$

where $C_{s_m} = \{C_{y_{s_i}}\}_{i=1}^{13}$, $C_{t_m} = \{C_{y_{t_i}}\}_{i=1}^{7}$ and $C_{r_m} = \{C_{y_{r_i}}\}_{i=1}^{6}$ are the cepstral coefficients derived from y_s, y_t and y_r, respectively, for each frame $m \in M$, where M is the total number of frames. The cepstral coefficients, C_m, is fed to a continuous density HMM for evaluating the word recognition performance.

6.7 The HMM word recognition with CN features

Continuous density HMM recognition was performed with isolated digits for male utterances from the TIDIGITS speech corpus. The HMM model parameters were the same as with the synaptic adaptation and the two-tone suppression, that is, HMM with 15 states per digit, 5 mixture components per state with diagonal covariances were used to define each model. A 3-state silence/pause model was inserted at the beginning of each utterance. There were 55 male speakers in the training set, each speaker with two utterances of the digits 1-9, 'oh' and 'zero' (total 1210 utterances in the training set) and and a separate set of 31 male speakers in the test set each speaker with two utterances of the digits 1-9, 'oh' and 'zero' (total 682 utterances in the test set). Test speech was corrupted with 4 different types of additive noise from the NOISEX 92 database: Gaussian white noise, factory noise, babble noise and Volvo noise. In the testing phase, a left-to-right Viterbi recognizer using word-level lattice network, a dictionary and a token passing algorithm was used for decoding. Table 6.1 shows the comparison in recognition rates between the base ZCPA and ZCPA with the proposed system based on CN processing strategy, (ZCPA_AUD), for the four types of additive noise.

TABLE 6.1: Comparison of recognition rates (%) between the base ZCPA, and ZCPA with CN processing strategy, (ZCPA_AUD), with isolated digits (TIDIGITS corpus in four types of additive noise).

SNR (dB)	White ZCPA	White ZCPA_AUD	Factory ZCPA	Factory ZCPA_AUD	Babble ZCPA	Babble ZCPA_AUD	Volvo ZCPA	Volvo ZCPA_AUD
Clean	95.4	100.0						
40	90.9	100.0	95.4	100.0	90.9	100.0	95.4	100.0
30	81.8	86.3	86.3	95.4	86.3	100.0	90.9	100.0
15	77.3	31.8	81.8	81.8	72.7	72.7	86.3	90.94
10	68.8	27.3	68.8	77.3	63.6	68.8	86.3	90.4
5	50.0	22.7	50.0	72.7	31.8	45.4	81.8	90.9

It is observed that the best performances with the proposed method are obtained with the Volvo noise, followed by factory noise, babble noise and white Gaussian noise. It is observed that in non-stationary factory, babble and Volvo noise, the performances are improved under all signal conditions and in all noise types. In wideband white Gaussian noise, the recognition rates with ZCPA_AUD are improved over the base ZCPA in clean condition and at high SNRs with a significant degradation at low SNRs. This is mainly due to the susceptibility of the mean discharge rate output to white noise, which is discussed in the next section.

The McNemar's significance test [102] was performed on the error rates from the two algorithms, ZCPA and ZCPA_AUD, with the test data consisting of 682 isolated digit utterances and it was found that the difference in the recognition results arising from the two algorithms were statistically significant.

It is observed from Table 6.1 that as in Table 3.3 and Table 4.3, many recognition rates are exactly equal for different SNRs and noise types, which was stated to be due to use of isolated digit recognition with male speakers only with a small number of models (eleven), and a very low vocabulary (eleven words) for the eleven digits with definite word boundaries. This produces a bias on misrecognition process. Usually, the misrecognised word is the same utterance across all speakers, such that the misrecognised words are always a fraction of 11, ie 10/11, 9/11, 8/11, etc. for one, two and three words misrecognised, respectively. The isolated words recognition has been used primarily to measure the relative performances of two algorithms, and the results have been tested for statistical significance.

6.8 Noise considerations in the mean rate (MR) processing

The average discharge rate in the rate-place representation is very sensitive to background noise, particularly to white noise, while the synchrony response is more immune to white noise than the mean discharge response [25].

The degradation of the mean discharge response in noise conditions is due to primarily two reasons. Firstly, the half-wave rectification process in the envelop detector introduces dominant harmonics and increases the statistical properties such as the mean and variance. Secondly, the high frequency perturbation of the white noise further accentuates the variance at the output.

The time-frequency plots of the MR output, y_r, for the utterance 'one' in clean condition and at 10 db SNR white noise are shown in Fig. 6.9. Each waveform is the output of one of the 16 channels, where the lower channels are higher frequency channels. It is observed from Fig. 6.9 that the high frequency segments are mostly affected while the low frequency segments are unaffected by the white noise perturbations.

The noise performance of the mean discharge response is also dependent on the time constant, τ, of the low-pass filter used for envelop detection. Fig. 6.10 shows the MR spectrogram for the male utterance of the digit 'one' in clean condition

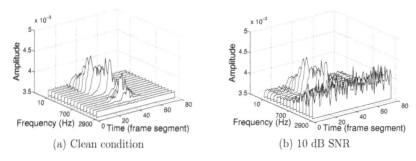

(a) Clean condition   (b) 10 dB SNR

FIGURE 6.9: The time-frequency plots of the MR output, y_r, for the utterance 'one' with and without white noise.

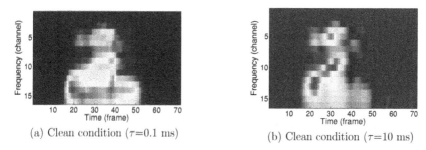

(a) Clean condition ($\tau=0.1$ ms)   (b) Clean condition ($\tau=10$ ms)

FIGURE 6.10: The MR spectrogram for the male utterance of the digit 'one' in clean condition for two time constants of (a) $\tau=0.1$ ms and (b) $\tau=10$ ms for the smoothing low pass filter. Lower number channels are higher frequencies. (In gray scale, lighter shade indicates higher intensity and darker shade implies lower intensity).

for the two time constants, a lower time constant of $\tau=0.1$ ms and a higher time constant $\tau=10$ ms. The same plots for 15 db SNR are shown in Fig. 6.11. It is observed from Fig. 6.10 (b) for clean condition that, at a higher time constant of $\tau=10$ ms, the high frequency components are enhanced. Moreover, in Fig. 6.11 (b) although it appears to contain more noise, the noise within the tonal segments are reduced at a higher time constant. That is, less noise is observed in the speech segments at the high frequency range. From the above analysis, it is inferred that a higher time constant used for MR feature extraction gives better performance.

One reason for the mean rate performance degradation in noise conditions is the increased variance in presence of additive noise. To further analyze this effect, we computed the unnormalized output data variance for $y_r(m, f)$ by summing the variances of all the time-frequency units for a digit utterance 'one' in clean and three values of SNR. The variance was measured for a low frequency channel at

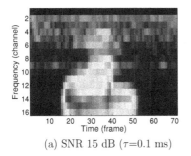

(a) SNR 15 dB (τ=0.1 ms)

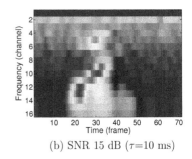

(b) SNR 15 dB (τ=10 ms)

FIGURE 6.11: The MR spectrogram for the male utterance of the digit 'one' with 15 dB SNR white noise for two time constants of (a) τ=0.1 ms and (b) τ=10 ms for the smoothing low pass filter. Lower number channels are higher frequencies. (In gray scale, lighter shade indicates higher intensity and darker shade implies lower intensity).

approximately 500 Hz and a high frequency channel at approximately 2500 Hz. A second order elliptic low-pass filter was used for envelope detection with a lower time constant of τ=0.1 ms and a higher time constant of τ=10 ms.

TABLE 6.2: Data variance of MR output, $y_r(m, f)$ (Fig. 6.8) for the utterance 'one' in white noise obtained with τ=0.1 ms and τ=10 ms at channel frequencies of 500 Hz and 2500 Hz. It is observed that the variance is reduced at a higher time constant for both high and low frequencies.

Noise (SNR,dB)	500 Hz		2500 Hz	
	$\tau = 0.1$ ms	$\tau = 10$ ms	$\tau = 0.1$ ms	$\tau = 10$ ms
Clean	4.5936	0.0240	2.46×10^{-4}	1.66×10^{-7}
30	4.6409	0.0240	0.0025	2.93×10^{-6}
15	4.1689	0.0235	0.1575	4.32×10^{-4}
5	3.0111	0.0203	0.3380	0.0027

The variances are shown in Table 6.2. It is observed that the variance decreases as the time constant is increased for both the high frequency channels and the low frequency channels. Although for experimental purpose only a single digit utterance 'one' was chosen, the trend is expected to be the same for other digit utterances, though the variance may differ. These results with MR processings are consistent with the results of Holmberg [27] that a high time constant in synaptic adaptation generally improves ASR performance.

A measure of noise susceptibility of the mean discharge response may be obtained from the coefficient of variation, defined as

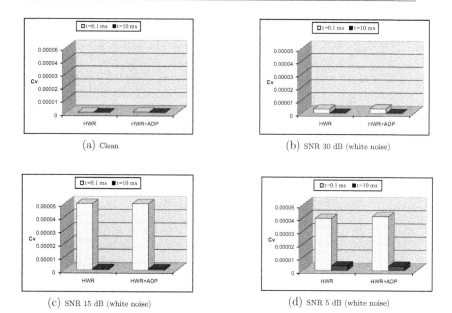

(a) Clean

(b) SNR 30 dB (white noise)

(c) SNR 15 dB (white noise)

(d) SNR 5 dB (white noise)

FIGURE 6.12: The coefficient of variation (C_v) for the MR processing, $y_r(m, f)$, for two time constants at the output of the FIR filter, at the output of the half-wave rectification (HWR) and at the output of the synaptic adaptation (HWR_ADP) for four different values of SNRs in Figs. (a)-(d). It is observed that the C_v values are reduced for higher time constants.

.

$$C_v = \text{Coefficent of variation} = \frac{\sigma_{MR}}{\mu_{MR}} \qquad (6.12)$$

where σ is the standard deviation and μ is the mean of $y_r(m, f)$, obtained by summing the variances and the means, respectively, of all the time-frequency units. It is noted from Eqn. (6.12) that a lower value of C_v should give a good immunity to noise.

The C_v values were computed at the normalized output of the half-wave rectifier (HWR) and the synaptic adaptation stage (HWR_ADP), which are defined in Fig. 6.8. Two time constants were used, a low time constant ($\tau = 0.1$ ms) and a high time constant ($\tau = 10$ ms). Measurements were taken in clean condition and three values of SNR at the output of a channel with a CF of approximately 2500 Hz with the isolated digit utterance 'one' as the input. C_v values are plotted in Fig. 6.12 in clean condition and three different values of SNRs in Figs. (a)-(d). It is observed

133

from Fig. 6.12 that the C_v values are lower at a higher time constant than at a lower time constant for all SNRs, which are consistent with the results of Figs. 6.10 and 6.11, and Table 6.2. A slight increase in the data variance is observed at the output of the synaptic adaptation over the variance at the HWR at the lower time constant, whereas a decrease in variance is observed at the higher time constant.

6.9 Performance comparisons between CN features and MFCC features

The MFCC is widely used for speech parametrization and is accepted as the standard for speech recognition applications and in research. We compared the performance of the base ZCPA with the base MFCC_0 in white and factory noise. Moreover, ZCPA_AUD was also compared with MFCC_0 with delta features (MFC_DEL), which was obtained by concatenating the delta features only (13 cepstra) with the base MFCC_0 using Eqn. (2.18) to obtain a 26 dimensional feature vector. MFC_DEL is restricted to 26-dimensions for an equivalent comparison with the 26 dimensional ZCPA_AUD. The MFCC features were obtained with a 25 ms frame window at 10 ms frame rate with 40 triangular filters (13 linear and 27 log) [129]. The comparison results are shown in Table 6.3.

The results indicate that ZCPA performs better than MFCC in white noise conditions while MFCC performs better than the ZCPA in non-Gaussian non-stationary factory noise. Similarly, ZCPA_AUD performs equal or better than MFC_DEL in white noise, but degrades in real-world factory noise. This is consistent with the results in [7] where it is stated that although the base ZCPA performs better than MFCC in noisy conditions, the relative performance improvement for the ZCPA in noisy conditions over the MFCC is greater in white Gaussian noise than in real-world noise, such as factory noise. It is also stated therein that the reason for this is not clear. One reason for this may be that in the ZCPA, the zero-crossing variance is reduced more for the high frequency white noise components than the lower frequency real-world noise (factory, car, and babble noise) components.

Therefore, although overall the ZCPA performs better in real-world noise than in white noise, the relative improvement in performance compared to MFCC is higher in white noise than in non-Gaussian real-world noise.

TABLE 6.3: Comparison of recognition rates (%) of ZCPA with MFCC_0, and ZCPA_AUD with MFCC_0 with delta features (26-dimensional MFC_DEL), in isolated TIDIGITS with male utterances in clean condition and in white and factory noise.

Noise (SNR) dB	White				Factory			
	ZCPA	MFCC	ZCPA_AUD	MFC_DEL	ZCPA	MFCC	ZCPA_AUD	MFC_DEL
Clean	95.4	100.0	100.0	100.0				
40	90.9	100.0	100.0	100.0	95.4	100.0	100.0	100.0
30	81.8	81.8	86.3	81.8	86.3	100.0	95.4	100.0
15	77.3	22.7	31.8	27.3	81.8	95.4	81.8	95.4
10	68.2	18.2	27.3	27.3	68.8	86.3	77.3	90.4
5	50.0	9.09	22.7	22.7	50.0	54.5	72.7	54.5

6.10   Summary and conclusions

In this chapter, a novel feature extraction method for ASR based on the processing strategy in the cochlear nucleus is proposed. Features were extracted by the parallel processing of three subsections corresponding to the AVCN, DCN and PVCN of the cochlear nucleus, each of which is assigned a particular task of detecting and enhancing the particular speech segments. Temporal place codes of the zero-crossings were used as a synchrony detector to discriminate the low frequency formants. The rate-place code as used in the auditory synapse were utilized for capturing the mean discharge rate corresponding to the envelope variations due to the input stimulus. Additionally, a companding method was utilized for spectral enhancement through two-tone suppression. It is shown that separating the synchrony detection from the synaptic processing, which is consistent with the morphological processing methodology in the CN, may improve ASR performance. It is further observed that a high time constant used for feature extraction from the synaptic processing may improve system performance in clean and white noise conditions.

The system was evaluated by HMM based recognition with features extracted from weighted cepstrum of the three processings using an isolated digits (TIDIGITS) speech corpus. The comparisons show improved recognition rates over the base ZCPA auditory model. The improvement is particularly significant in clean condition, which is due to the processing strategy of the CN employed, and maybe due to the added processing of the mean rate and the two-tone suppression. These have the effects of further enhancing speech segments for better recognition performances. In noisy conditions, improvements are observed in non-Gaussian real-world noise at all SNRs. However, there is a degradation in white Gaussian noise at low SNRs, mainly due to the susceptibility of the mean discharge response to white Gaussian noise.

Chapter 7

Psychoacoustic spectral subtraction methods for speech recognition

In this chapter, a noise suppression method for speech enhancement based on the psychoacoustics of auditory perception is proposed. The traditional approach for speech enhancement is spectral subtraction in which the estimate of the noise power spectrum, obtained in non-speech intervals of the signal is subtracted from the the short-time spectral amplitudes of the degraded signal, based on the assumption that speech and additive noise degradation are uncorrelated. A variety of criteria may be used for obtaining an optimal statistical operator for the subtraction algorithm based either on signal spectral variance or on minimum mean square error estimation (MMSE) [134], and using various Wiener filtering. The effectiveness of the spectral subtraction relies primarily on the consistency and accuracy of noise magnitude spectrum estimates. This method has two important drawbacks. Firstly, a speech detector is required to determine intervals from which a reliable noise estimate can be obtained, and secondly, the estimate of the noise magnitude spectrum is blindly subtracted from the magnitude spectrum of the noisy speech segments.

High variance of the noise magnitude spectrum estimates may result in under-subtraction, in which there is some residual noise not being subtracted, which produces sharp spurious random peaks in the magnitude spectrum of the enhanced speech. This artefact is commonly referred to as "musical noise", and is the major disadvantage of the spectral subtraction method. Moreover, strong over-subtraction results in a reduction in the speech content, or may result in negative power spectral values. Various methods have been proposed to reduce the effects of residual noise such as magnitude averaging [131], oversubtraction of noise and introduction of a

137

spectral floor [132], soft-decision noise suppression filtering [133], optimal MMSE estimation of short-time spectral amplitude [134], nonlinear spectral subtraction [135], and introduction of morphological-based spectral constraints [136]. Unfortunately, these methods are not effective at a very low SNRs. The main reason is the difficulty to suppress noise without decreasing intelligibility or introducing distortions of the tonal components.

The state-of-the art speech enhancement method by Martin [137] utilizes a noise power spectral density estimation based on optimal smoothing and minimum statistics. It tracks the spectral minima in each frequency band, and by minimizing a conditional mean square estimation error criterion, an optimal smoothing parameter is obtained which is used for for recursive smoothing of the power spectral density of the noisy speech signal. Based on this optimally smoothed power spectral density estimate and the analysis of the statistics of the spectral minima, an unbiased noise estimator is obtained.

Due to the immunity of the human auditory system to noise, a recent trend has been the use of psychoacoustics for speech enhancement. Psychoacoustic models have been used successfully in audio coding to reduce bit rates and for audio compression. The MPEG1 Layers I, II and III audio compression standards are the state of the art in the coding of audio signals. These coding standards achieve remarkable compression by exploiting two psychoacoustic characteristics of the human auditory system, that is, the critical band resolution and the auditory masking property [138].

In the past few decades, psychoacoustics have been used for speech enhancement and improvement of objective quality of speech and speech intelligibility [139], [140]. Schroeder *et al.* [57] utilized the property of auditory masking for measuring the perceptual degradation of speech signal by the quantization noise which arises in audio coding. It was based on the assumption that the subjective loudness of noise is reduced by the presence of a masking signal in the speech. The method consisted of a critical band analysis followed by frequency masking implemented by a spreading function for the speech and the noise determined separately. The degradation in speech was measured by the relative loudness of the noise in the presence of speech signal above a minimum masking threshold and was used as an objective measure of speech quality for optimizing the performance of digital speech coders. The psychoacoustic method proposed by Tsoukalas *et al.* [141] utilized an auditory masked threshold (AMT) of the noise across the frequency spectrum to adjust the parameters in the subtraction process. The method improved the speech intelligibility by 40% as determined by an objective listening test. The method proposed by Arehart *et al.* [142] was based on the AMT method similar to Schroeder *et al.* [57]. A noise subtraction algorithm was implemented based on the AMT which was evaluated for speech intelligibility in the hearing-impaired listeners and was shown to improve performance.

Lai *et al.* [143] proposed a psychoacoustic spectral subtraction method based on the auditory masking property to reduce the artifacts in a conventional spectral subtraction utilizing a voice activity detector. It determined an audible threshold obtained from a psychoacoustic perceptual model used in the MPEG-1 audio coding. The subtraction rule followed in the method was that it was made smaller than the audible threshold. The method proposed by Shao *et al.* [144] utilized the masking properties of the human auditory system. The method consisted of thresholding wavelet transform coefficients relative to a noise masking threshold which was computed by modelling the frequency selectivity of the human ear and its masking property. The parameters of the time-frequency subtraction algorithm were adaptively adjusted to the masking threshold to reduce background and residual noise and distortions.

In this chapter, a time-frequency spectral subtraction method based on the peripheral auditory system is proposed. The method has four distinctive features compared to the above methods. Firstly, it utilizes several psychoacoustic properties of human perception, such as the critical band filtering, synaptic adaptation which also introduces temporal forward masking, equal loudness pre-emphasis, power law compression of hearing, and simultaneous masking effect introduced by a spreading function. The processing is similar to the PLP processing of speech with added perceptual features of adaptation and simultaneous masking effects. Based on these effects, a minimum masking threshold similar to the AMT is determined following the methods of Schroeder *et al.* [57]. Hence, the proposed method determines the AMT considering a more detailed perceptual transformation that occur in the peripheral auditory system.

In noisy speech, noise may be masked by the much stronger tonal speech components. This was verified in [57] by perceptual experiments which indicated that the loudness of noise was reduced in presence of a masking signal. We also estimate separately the noise after it undergoes the psychoacoustic transformations as mentioned above for the tonal components, which we call the perceptual noise.

Secondly, spectral subtraction methods in [141], [144] use simultaneous masking properties of the human auditory system for estimating a masking threshold. Most recently, noise suppression using both simultaneous masking and the temporal masking using fractional gammatone filters have been reported [145], [94]. It was also reported that temporal masking thresholds are less susceptible to noise than the simultaneous masking thresholds. The method proposed in this chapter utilizes both simultaneous and temporal masking, but implements the auditory synaptic adaptation to simulate forward temporal masking.

Thirdly, the performance of speech enhancement methods are usually evaluated by some objective measures of speech quality improvement such as SNR improvement, listening tests (the mean opinion score (MOS) and the dynamic rhyme test (DRT), nonsense syllable test, hearing-in-noise, etc.) or by other perceptual measures. In these methods, psychoacoustics are tested on human subjects for evaluating improvements in speech intelligibility. However, we evaluated the performance by a speech recognition process, which is, in the case of noisy speech, an indirect effect of SNR improvement. Particularly, the spectral subtraction method is compared with the PLP processing of speech for evaluating its relative performance in speech recognition. In recent studies, speech recognition thresholds have been related to the speech-to-noise ratio required to achieve a particular level of intelligibility (usually 50%) [142].

Fourthly, the AMT approach [142] to speech enhancement assumes that some components of the noise will be masked (below threshold) by the speech. Any noise components that are below this masked threshold will not be detectable by the human listener and so perceptually are not important components to suppress. The objective, then, is to minimize only the audible portion of the noise spectrum. This is particularly true when the enhanced speech is tested on human subjects. Additionally, we show that, for ASR applications, further improvements in recognition performance may be obtained by augmenting the masking of the noise by the AMT by spectral subtraction in the masked region. The strategy is to remove the masked noise from the ASR system, similar to the masking effect in the human auditory system.

Based on the AMT, and the estimated perceptual noise, we have implemented two spectral subtraction algorithms: a straight-forward scheme of subtracting the total estimated perceptual noise from the noisy speech spectrum, which we refer to as subtraction without thresholding, and a spectral subtraction of the noise which lies below the masking threshold, which we refer to as subtraction with thresholding (AMT). It was observed that, both methods give significant improvements over the base PLP performance, with the subtraction with thresholding method giving better recognition results.

7.1 Computation of the auditory masking threshold (AMT)

7.1.1 The power spectrum

Assuming that speech and noise are uncorrelated, we may express the noisy speech, $s(n)$, as

$$s(n) = m_s(n) + d(n) \tag{7.1}$$

where $m_s(n)$ and $d(n)$ represent the clean speech and the additive white Gaussian noise with zero mean, respectively.

If $w(n)$ is a window of finite length, N, ending at time m, then the value of $s(n)$ at the frame index m is given by $s(n;m) = s(n)w(m-n)$. Using discrete short-time Fourier transform (STFT) of $s(n;m)$, we obtain from Eqn. (7.1)

$$\Gamma_s(m,\omega) = \Gamma_{m_s}(m,\omega) + \Gamma_d(m,\omega) \tag{7.2}$$

where $\Gamma_s(m,\omega)$, $\Gamma_{m_s}(m,\omega)$, and $\Gamma_d(m,\omega)$ are the STFT of the noisy speech, the clean speech, and the noise, respectively, at the frame index, m. The magnitude spectrum of clean speech can be estimated using the generalized expression for spectral subtraction [130]

$$
\begin{aligned}
|\hat{\Gamma}_{m_s}(m,\omega)| &= (|\Gamma_s(m,\omega)|^\alpha - \beta|\hat{\Gamma}_d(m,\omega)|^\alpha)^{\frac{1}{\alpha}}, \text{if } (|\Gamma_s(m,\omega)|^\alpha - \beta|\hat{\Gamma}_d(m,\omega)|^\alpha)^{\frac{1}{\alpha}} > \gamma|\hat{\Gamma}_d(m,\omega)| \\
&= \gamma|\hat{\Gamma}_d(m,\omega)|, \qquad\qquad\qquad \text{otherwise,}
\end{aligned}
\tag{7.3}
$$

where α, β, γ are positive constants, and $|\hat{\Gamma}_d(\omega)|$ is an estimate of the noise magnitude spectrum (periodogram). The parameter α determines the subtraction domain (if $\alpha=1$, subtraction is performed in magnitude spectral domain and if $\alpha=2$, the subtraction occurs in the power spectral domain). The parameter β controls the amount of over-subtraction [132]. Strong over-subtraction results in strong noise suppression, however, at the same time there is a reduction in the speech content. On the other hand, lower values of β result in suppression of additive background noise but introduces very audible musical noise. The parameter γ is used to prevent from producing negative spectral values, by limiting these to the spectral floor. This reduces dynamic the range of the enhanced speech and, hence also lowers the dynamic range of the spurious spectral peaks. As a result, the intensity of musical noise is also reduced.

The psychoacoustic transformation which the input noisy signal $s(n)$ undergoes in the peripheral auditory system is shown in Fig. 7.1. The power spectrum of the noisy speech is defined as the square of the magnitude spectrum and can be written as

$$S(n,\omega) = \text{Re}[\Gamma_s(n,\omega)]^2 + \text{Im}[\Gamma_s(n,\omega)]^2 \qquad (7.4)$$

and the power spectrum (periodogram) for the noise as

$$D(n,\omega) = \frac{1}{K}\left(\text{Re}[\Gamma_d(n,\omega)]^2 + \text{Im}[\Gamma_d(n,\omega)]^2\right). \qquad (7.5)$$

where K is the length over which the noise is measured.

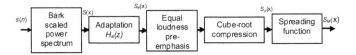

FIGURE 7.1: The psychoacoustic transformations of the input speech.

7.1.2 The critical band analysis

The FFT frequencies f are expressed in terms of the perceptual critical band numbers x ($x=1,\ldots,23$) using the transformation [57]

$$f = 650\sinh\left(\frac{x}{7}\right). \qquad (7.6)$$

x may also be expressed in mm as distance along the length of the BM as a function of the tuning frequency with values corresponding to the critical band (characteristic) frequencies along its length. Thus, the "best" frequency is approximately 1 kHz when the distance along the BM is approximately $x=8.5$ mm. Analytically, x in mm may be determined from Eqn. (2.1) by first determining the normalized x (between 0 and 1) and then unnormalizing by using the fact that $x=1$ when BM length is 32mm.

The power spectrum $S(n,\omega)$ is expressed in terms of the critical band numbers x, $x=1,\ldots,23$, by combining a 256-point FFT into 23 Bark bins using the Bark filter coefficients. The filter coefficients, given by $M_x(n)$, $n=1,\ldots,N$, are obtained from a 128x23 matrix (corresponding to FFT/2 points and 23 Bark filters) of weights which are used to combine the FFT/2 bins into 23 Bark bins [129].

$$S(x) = \sum_{k=0}^{N/2} S(k,\omega)M_x(k) \qquad x = 1,\ldots,23. \tag{7.7}$$

7.1.3 Adaptation (temporal forward masking)

After critical band filtering, the output was processed by a temporal synaptic adaptation stage as described in Sec. 3.1.1, whereby the rapid and short-term synaptic adaptation was implemented by a first order high-pass IIR filter function with a time constant of 250 ms (Eqn. (3.22)). Forward masking can be viewed as a consequence of auditory adaptation [24]. It is observed in Fig. 3.3 that synaptic adaptation may produce effects similar to forward masking. For example, it is seen that the first tone burst is reduced in magnitude due to the presence of the second tone following it when the time duration separating the two tones is less than the time constant of the filter. However, for larger time constants, where the frame size is less than the synaptic adaptation time constant, the forward masking effects may depend on the previous frames [94]. However, in the proposed method the effects of synaptic adaptation of the previous frame have not been considered.

The synaptic adaptation filtering was implemented in the time domain by spectral inversion of the power spectrum after the critical band binning. In spectral inversion, the binned FFT magnitudes are inverse Fourier transformed, the adaptation filtering is applied in the time domain data, and the output is again transformed to the FFT magnitudes. It is expected that with a frame size of 25 ms, the effects of rapid synaptic adaptation in the initial 3 ms of the onset would be emphasised since the synaptic adaptation filtering are performed in the time domain by spectral inversion. The synaptic adaptation filtering is given as

$$S_a(z) = H_a(z)S(z) \tag{7.8}$$

where $H_a(z)$ is the transfer function of the synaptic adaptation filter given by Eqn. (3.22). After the adaptation filtering, the output, $S_a(z)$, was converted back to the power spectral domain, $S_a(x)$.

The effect of adaptation filtering using Eqn. (7.8) is illustrated in Fig. 7.2 (a) where it is shown that with synaptic adaptation the masking threshold in dB is slightly raised over when synaptic adaptation is not used. In Fig. 7.2 (b) it is shown in linear scale that with the synaptic adaptation filtering, the masking threshold is raised above the base PLP utilizing no synaptic adaptation filtering. Further explanation is given in Sec. 7.1.6.

7.1.4 Equal loudness pre-emphasis and cube-root compression

The synaptic adaptation output, $S_a(x)$, was transformed by equal loudness pre-emphasis $E(\omega)$ given in (Eqn. (2.17)) and then compressed by taking the cube root to simulate the power law of hearing, as is done in the conventional PLP processing [70]. The transformed spectrum $S_p(x)$ is the audible spectrum according to the transformations occurring in the peripheral auditory system, and is given by

$$S_p(x) = [S_a(x)E(\omega)]^{0.33} \tag{7.9}$$

7.1.5 The spreading function

Frequency masking does not only occur within the critical bands, but also spreads to the neighboring bands. This was simulated by a spreading function, B, given in [57] as

$$10\log B(x_i, x_j) = 15.81 + 7.5(x_i - x_j + 0.474) - 17.5\sqrt{(1 + (x_i - x_j + 0.474)^2} \tag{7.10}$$

where x_i and x_j, $i=1,\ldots,23$, and $j=1,\ldots,23$, are the Bark frequencies corresponding to the masked signal and the masking signal, respectively. The spreading function has lower and upper skirts with slopes of +25 dB and -10 dB per critical band. The masking ability of a given signal component depends on its frequency position and its loudness. A signal or noise which is farther away in frequency than a masker signal will be less masked than a signal or noise nearer to it.

The spectrum, $S_p(x)$, for each data frame was converted to an excitation (masking) pattern, $S_M(x)$, by multiplying (equivalent to a convolution in the time domain) with the BM spreading function $\mathbf{B}$, given in Eqn. (7.10),

$$S_M(x) = S_p(x) \times \mathbf{B}, \qquad x = 1,\ldots,23 \tag{7.11}$$

where $S_M(x)$ and $S_p(x)$ are 1×23 matrices for each frame data and $\mathbf{B}$ is a 23×23 matrix given by

$$\mathbf{B} = \{b_{i,j}\} = \begin{pmatrix} b_{1,1} & b_{1,2} & \cdots & b_{1,23} \\ b_{2,1} & b_{2,2} & \cdots & b_{2,23} \\ \cdot & \cdot & \cdots & \\ b_{23,1} & b_{23,2} & \cdots & b_{23,23} \end{pmatrix}$$

where b_{ij} is the value of the spreading function at the i-th filter due to the effect of the j-th filter.

The excitation pattern, $S_M(x)$, can be thought of as energy distributions along the basilar membrane and may be considered as the effects of frequency masking as the energy spreads along the BM [57].

7.1.6   The auditory masking threshold (AMT)

As a consequence of the synaptic adaptation and simultaneous masking effects due to the spreading function, an auditory masking threshold, $T(x)$, was determined. The masking threshold was found by multiplying the excitation function, $S_M(x)$, by a sensitivity function, $w(x)$,

$$T(x) = S_M(x) \times w(x), \qquad x = 1, \ldots, 23 \tag{7.12}$$

where the sensitivity function, $w(x)$, was given in [57] as

$$10 \log w(x) = -(15.581 + x) \quad \text{dB} \tag{7.13}$$

such that for $f=1$ kHz (corresponding to a distance of $x=8.5$ mm along the BM), the masked threshold, $T(x)$, is approximately 24 dB below the speech excitation pattern, $S_M(x)$. As signal may mask noise, noise may also have a masking effect on the signal. Since the excitation pattern was obtained from speech corrupted with noise, both effects are reflected in the masking threshold thus computed. If the noise signal lies below the masking threshold, then it would be inaudible.

Fig. 7.2 (a) shows the excitation pattern, $S_M(x)$, to a pure tone centered at 1 kHz and the masking threshold, $T(x)$, without and with synaptic adaptation. The output was obtained by summing the power spectrum outputs for all the critical bands for all frames, converted to dB, and plotted across the critical band frequencies. For reference, the threshold in quiet is also plotted on the same graph. The slope of the masking threshold is steeper towards lower frequencies, indicating that higher frequencies were more easily masked. It is also observed in Fig. 7.2 (a) that with synaptic adaptation, the masking threshold is slightly raised (in dB scale), indicating increased masking effects with synaptic adaptation over the conventional PLP processing [70], as explained in Sec. 2.6.2. This effect is also observed in Fig. 7.2 (b) which shows the masking threshold in a linear scale instead of a dB scale for both synaptic adaptation (temporal forward masking) and the simultaneous (spreading) masking, computed separately. It is observed that simultaneous masking has a greater effect on the masking threshold than the temporal forward masking due to the auditory adaptation.

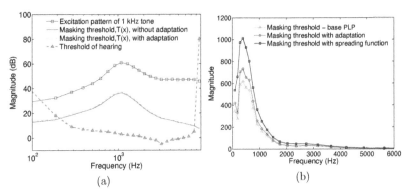

(a)   (b)

FIGURE 7.2: (a) The excitation pattern, $S_M(x)$ and the masking thresholds, $T(x)$, to a tone burst centered at 1 kHz without and with synaptic adaptation in dB scale (y-axis) and Bark frequency scale (x-axis). (b) The comparison of the masking thresholds, $T(x)$ with synaptic adaptation and with spreading function, shown in a linear magnitude-frequency scale. It is observed in (a) that with synaptic adaptation, the masking level is slightly raised.

7.2 Estimation of the perceptual noise

For a spectral subtraction algorithm, a reliable estimate of the noise is required. In the human auditory system, noise undergoes the same psychophysical transformations as for the tonal speech. To estimate the noise after perceptual transformations, the noise was first independently determined from the input speech from the average of the first seven frames which are considered as the silent/pause segments. Next, the power spectral density, $D(\omega)$, was estimated by a periodogram, using a N-point FFT. This was filtered by the critical band filters by multiplying with the Bark filter coefficients, $M_x(n)$, n=1,...,N,

$$D(x) = \sum_{k=0}^{N/2} D(k,\omega)M_x(k), \qquad x = 1,\ldots,23. \qquad (7.14)$$

The temporal IIR adaptation filtering was next performed, followed by equal loudness pre-emphasis and the cube root compression, as done for the tonal speech in Sec. 7.1.3 and 7.1.4, to obtain $D_p(x)$. The estimated perceptual noise, $D_M(x)$, after the effects of spreading to the adjacent filters in the filterbank, was computed by multiplying $D_p(x)$ with the spreading function, **B**, given in Eqn. (7.10),

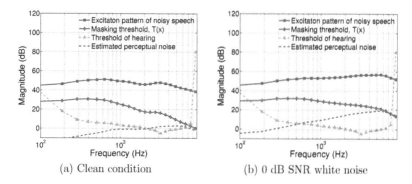

(a) Clean condition   (b) 0 dB SNR white noise

FIGURE 7.3: The excitation pattern of input speech, $S_M(x)$, the masking thresholds, $T(x)$, and the estimated perceptual noise, $D_M(x)$ for an utterance 'one' in (a) clean conditions and (b) 0 dB SNR white noise. It is observed in (b) that white noise has the effect of raising the masking threshold, particulary at higher frequencies, whereas the effect of perceptual noise is negligible in clean conditions.

$$D_M(x) = D_p(x) \times \mathbf{B}, \qquad x = 1, \dots, 23. \qquad (7.15)$$

where $D_M(x)$ and $D_p(x)$ are 1×23 matrices representing each data frame and $\mathbf{B}$ is a 23×23 matrix.

Fig. 7.3 shows the excitation pattern of the input speech, $S_M(x)$, the masking threshold, $T(x)$, and the estimated perceptual noise, $D_M(x)$, for an utterance 'one' in clean conditions (a) and at 0 dB SNR white noise (b). The output along the y-axis was obtained by summing the power spectrum magnitudes for all the critical bands for all frames, then converted to dB, and plotted across the critical band frequencies (along the x-axis). For the whole utterance, the spectrum may vary from phoneme to phoneme or segment to segment, thus the summing the frames will have a smoothing effect on the resultant spectrum. However, since the primary objective is to compare the effects of white noise, the spectrum smoothing effect, if any, would not be significant. Ideally, perceptual noise should be nil in clean condition. However, zero-noise condition in real life are not practical. Since in our case, the algorithm is equally used in both conditions, a small noise has been shown. For reference, the threshold in quiet is also plotted on the same graph. It is observed in (a) that in clean conditions, the perceptual noise has virtually no effects, whereas in (b) the noise floor is raised at higher frequencies in white noise. The spectrum of the corrupted speech is also raised, which raises the masking threshold (AMT) due to the presence of noise.

At low level noise, all of the the perceptual noise, $D_M(m, x)$, (m being the frame index) lies below the masking threshold level, $T(x)$. As the noise increases, the noise floor will exceed the masking threshold at points depending on the signal intensity of the tonal components. However, it is observed that at a very low SNR, this effect becomes insensitive to the increasing noise levels and any further increase in noise do not produce corresponding crossings of the masking level. This effect is shown in Table 7.1 for a digit utterance 'one'. It is expected that the values will not change significantly for other digit utterance such as 'two' and the trend will be similar. Therefore, only a representative digit has been chosen, so that a consistency is maintained and the results may be compared where applicable.

TABLE 7.1: Percentage of T-F cells in perceptual noise which are masked by the masking threshold for an utterance 'one' in white noise.

Noise SNR (dB)	Number of T-F cells masked by the masking threshold	%
Clean	1771	100.00
40	1626	91.81
30	1584	89.44
20	1569	88.59
10	1569	88.59
0	1569	88.59

This saturation effect may be due to the fact that at a high noise, the masking of the tonal components by the noise plays an increasing role, which raises the masking level in equal proportion to the increase in the perceptual noise level. Therefore, the extent of noise which is masked by the masking threshold saturates with increasing noise level. This effect may be related to the nonlinearity associated with the human masking property.

In Fig. 7.4, the masking threshold, $T(x)$, for an utterance 'one' in white noise is plotted as a function of the SNR. It is seen that as the noise increases, the threshold increase is small above 20 dB SNR, but increases significantly below 20 dB SNR, which is consistent with the data in Table 7.1.

7.3 The T-F noise subtraction algorithm based on the AMT

Fig. 7.5 shows the schematic of the psychoacoustic model for spectral subtraction using the auditory masking threshold. For testing purpose, the posterior SNR was

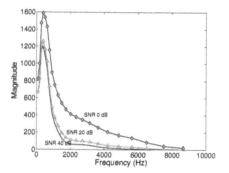

FIGURE 7.4: The masking threshold, $T(x)$, as a function of the SNR for an utterance 'one' in white noise. It is seen that the threshold is raised in presence of noise, the increase being significantly higher below 20 dB SNR.

determined from estimates of the clean speech and the noise (assuming white Gaussian noise) and expressed as the ratio of signal power to noise power in decibels. Clean speech, $m_s(n)$, was estimated from the noisy signal by using Martin's method of optimal smoothing and minimum statistics [137]. Input noise $\hat{d}(n)$ was estimated from the average of the first seven frames which were the silence/pause segments. The posterior SNR was calculated as

$$ \mathrm{SNR}_p = 10 \log \left(\frac{\sigma_{m_s}^2}{\frac{1}{k} \sum_k A^2 \sigma_d^2} \right) \tag{7.16} $$

where $\sigma_{m_s}^2$ and σ_d^2 are the variances of the estimated clean speech and noise, respectively, and k is the number of frames used to estimate the noise.

Based on the estimated perceptual noise, several spectral subtraction strategies may be implemented. A straight-forward scheme would simply subtract the estimated noise from the noisy speech. We implemented two time-frequency spectral subtraction algorithms to compare the relative effects.

1. For a given SNR_p, the total estimated perceptual noise, $D_M(x)$, was subtracted from the excitation pattern of the corrupted speech, $S_M(x)$. The clean speech spectrum, $M_s(m, x)$, was estimated using the following subtraction rule

$$ \hat{M}_s(m, x) = S_M(m, x) - D_M(m, x) \tag{7.17} $$

where m is the frame index and x is the frequency index in Bark number.

2. We utilized the masking threshold (AMT) to impose a constraint that the

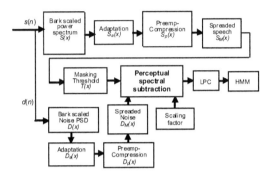

FIGURE 7.5: Schematic of the psychoacoustic model for spectral subtraction using the auditory masking threshold (AMT), $T(x)$.

perceptual noise which lay below the masking threshold was masked, and may be subtracted from the the excitation pattern of the corrupted speech spectrum, $S_M(x)$. The strategy is to remove the masked noise from the ASR system, similar to the masking effect in the human auditory system. For each time-frequency unit, the perceptual noise spectrum is compared with the masking threshold spectrum. If the perceptual noise is below the minimum masking level, then the estimated perceptual noise is subtracted from the excitation pattern of the corrupted speech spectrum. The clean speech spectrum, $M_s(m, x)$, is estimated using the following subtraction rule

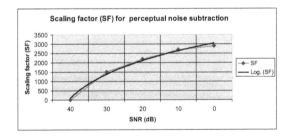

FIGURE 7.6: Noise scaling factor (SF) for the spectral subtraction utilizing the AMT as a function of SNR_p. The plot is shown with a logarithmic curve fitted to it.

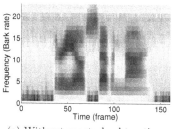
(a) Without spectral subtraction

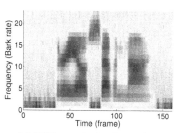
(b) With spectral subtraction

FIGURE 7.7: Spectrogram of the utterance '3o79' with 20 dB SNR white noise (a) without spectral subtraction, and (b) with perceptual spectral subtraction utilizing the AMT method. (In gray scale, a lighter shade indicates lower intensity and a darker shade implies higher intensity.)

$$\hat{M}_s(m,x) = S_M(m,x) - D_M(m,x) \times \text{SF} \qquad \text{if } D_M(m,x) < T(m,x) \qquad (7.18)$$

The scaling factor, SF, is like a gain function for the subtraction noise and is a function of the posterior SNR. The SF factor is multiplied to the perceptual noise spectrum before subtraction from the noisy speech spectrum. For a particular SNR, recognition run was conducted with several scaling factors until the optimum scaling factor was obtained for the best recognition result at that SNR. Fig. 7.6 was constructed from such optimum values of the scaling factor for SNR 0 to 40. The plot is shown with a logarithmic curve fitted to it. The SF also shows similar saturation nonlinearity corresponding to the data in the Table 7.1.

An autocorrelation matrix was obtained from the resulting power spectrum, which was solved for the LPC coefficients using the Levinson recursion, as is done in the PLP processing. The LPC coefficient matrix was converted to the spectral and then to the cepstral domain to obtain 13 cepstra per frame. This was input to a HMM recognizer for evaluation of the the the effectiveness of the psychoacoustic speech enhancement technique in terms of SNR improvements.

Fig. 7.7 shows the spectrogram of the utterance '3079' in 20 dB SNR white noise with no spectral subtraction and with perceptual noise subtraction utilizing the AMT in (a) and (b), respectively. It is observed in (b) that the noise is suppressed without affecting the tonal components.

Fig. 7.8 shows the effects of noise suppression utilizing the AMT across the frames (along the x-axis) for the utterance '3o79' in 20 dB SNR white noise. The

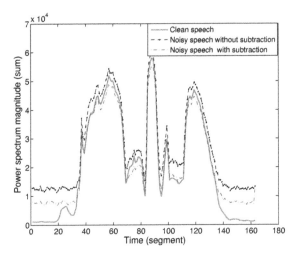

FIGURE 7.8: The effects of the perceptual spectral subtraction utilizing the AMT for the utterance '3079' in 20 dB SNR white noise across the frames (along the x-axis)

output, shown along the y-axis, was obtained by summing the power spectrum magnitudes for all the critical bands for each frame. It is observed from the figure that with the application of the perceptual noise subtraction utilizing the AMT, the noise is significantly reduced from the noisy speech spectrum as compared with the clean speech.

Fig. 7.9 shows the effects of noise suppression across the frequencies for the utterance '3o79' in 20 dB SNR white noise utilizing the AMT. The output, shown along the y-axis, was obtained by summing the spectrum magnitudes for all the frames for each critical band. It is observed that with the perceptual spectral subtraction, the noise in the high frequency is suppressed without any under-subtraction or over-subtraction. The summing of all the spectrum magnitudes ignores the finer variations of the magnitudes across the frequency range. However, the main purpose of the speech enhancement experiment is to compare the effects of noise for the three cases of clean condition, with perceptual spectral subtraction, and without perceptual spectral subtraction.

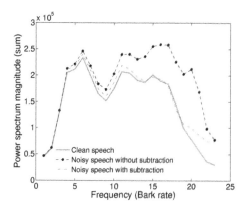

FIGURE 7.9: The effects of the perceptual spectral subtraction utilizing the AMT for the utterance '3079' in 20 dB SNR white noise as a function of frequency (along the x-axis)

7.4   HMM word recognition with psychoacoustic spectral subtraction

Speaker independent isolated digits from the TIDIGITS speech corpus were used for recognition experiments. There were 55 male speakers in the training set, each speaker with 2 utterances of the digits 1-9, 'oh' and 'zero' (total 1210 utterances in the training set) and a separate set of 31 male speakers in the test set each speaker with 2 utterances of the digits 1-9, 'oh' and 'zero' (total 682 utterances in the test set). The HMM model parameters and the Viterbi test parameters were the same as in the previous isolated word recognition, that is, diagonal covariance were used in the HMM with 15 states, 5 Gaussian mixture components per state and a token passing algorithm with dynamic programming was used for decoding. Test speech was corrupted with additive Gaussian white noise from the NOISEX 92 database.

In Table 7.2, the recognition results of the perceptual spectral subtraction using the two methods (without thresholding and with thresholding (AMT)) were compared with two other algorithms, the baseline PLP processing [70] and the perceptual spectral subtraction with optimal smoothing and minimum statistics [137] both of which were implemented experimentally in this research.

The baseline PLP features were obtained by using the same procedure and parameters as the perceptual spectral subtraction method described above, but without the perceptual processings. The recognition rates for the PLP were determined for different orders of LPC coefficients ranging from 10 to 20. The optimal value

TABLE 7.2: Word Recognition rates (%) of conventional PLP processing compared with psychoacoustic spectral subtraction without and with the masking threshold for isolated digits and white additive noise.

SNR (dB)	Baseline PLP	PLP with spectral subtraction with optimal smoothing and min. statistics	Perceptual spectral subtraction without thresholding	Perceptual spectral subtraction with AMT thresholding
Clean	100.0	100.0	100.0	100.0
40	100.0	100.0	100.0	100.0
30	81.8	86.4	86.4	95.4
20	27.3	45.4	77.2	90.1
10	18.2	22.7	63.6	63.6
0	9.0	13.6	31.8	18.2

was found to be 14. The spectral subtraction using minimum statistics were implemented in the PLP using the subtraction algorithm in [137] and the algorithm provided by the authors on the web. Thus the three algorithms PLP, PLP with spectral subtraction using the minimum statistics, PLP with spectral subtraction using psychoacoustics could be compared on the same platform and utilizing the same benchmarks.

It is observed that the perceptual spectral subtraction method demonstrated significant improvements in recognition performance over the baseline PLP performance and the minimum statistics method. However, the method which subtracts the total perceptual noise without reference to the masking threshold (column 4) performed worse than the method in which only the noise which is below the masking threshold is subtracted from the noisy speech (column 5), except at very low SNRs. For example, at 20 dB SNR, the recognition rate with the base PLP is 27.3%, and with spectral subtraction with minimum statistics is 45.4%, 77.2% with total noise subtraction without thresholding, whereas a recognition rate of 90.1% was achieved with perceptual spectral subtraction with the thresholding. The degradation when the total noise is subtracted may be due to the effects of oversubtraction which affects the high frequency tonal components. This may be seen from the time-frequency plot of the masking threshold and the perceptual noise in Fig. 7.10 where it is observed that the noise exceeds the masking threshold in the high frequency regions (corresponding to higher channels). Here the high frequency tonal components refer to high frequency speech components related to articulation, and not the noise artefacts.

However, at very low SNRs (below 10 dB SNR), a degradation of the performance with thresholding is observed. This is due to the fact that the extent of noise which is masked by the masking threshold saturates with increasing noise, as seen from Table 7.1. Hence, at very low SNRs, the substantially increased noise level above the threshold degrades the performance. Hence, a total subtraction of perceptual noise performs better than when only the noise below the threshold is subtracted.

The experiments were conducted in white noise only because the main objective is to compare the proposed perceptual speech enhancement method with other algorithm such as the PLP and PLP with spectral subtraction. White noise has spectral components at all frequencies whereas, real-world noise contains primarily low frequency components.

To test the significance of the difference in the error rates between the base PLP and the perceptual spectral subtraction utilizing the masking threshold, we applied the McNemar's significance test [102] (Appendix A) on the test set of 682 isolated digit utterances corresponding to the recognition results for 30 dB, 20 dB, 10 dB and 0 dB. The test results indicated that the differences in error rates due to the two algorithms are statistically significant.

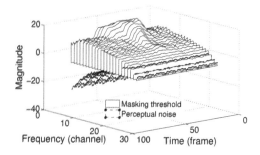

FIGURE 7.10: The masking threshold and the perceptual noise for the utterance 'one' in 20 dB SNR white noise. It is seen that the noise exceeds the masking threshold at higher frequencies, corresponding to higher channels. The magnitude along the z-axis is shown in dB scale.

7.5   Summary and conclusions

In Chapter 7, a time-frequency spectral subtraction method based on several psychoacoustic properties of human perception is proposed for evaluation in an ASR front-end. These properties are critical band filtering, synaptic adaptation which also introduces temporal forward masking, equal loudness pre-emphasis, power law compression of hearing, and simultaneous masking effect introduced by a spreading function. Based on these psychoacoustic transformations, a tonal excitation pattern is determined for speech corrupted with white noise. An auditory masking threshold (AMT) is computed from the tonal excitation pattern based on these effects. Noise also undergoes the same perceptual transformations and is separately estimated based on these perceptual features. It is observed that both synaptic adaptation and the simultaneous masking raise the masking threshold, with the simultaneous masking having greater effects.

The perceptual spectral subtraction method is evaluated by a speech recognition process for measuring performances by SNR improvements. Two time-frequency spectral subtraction methods are implemented: total noise spectral subtraction without reference to the masking threshold, and spectral subtraction in the region below the masking threshold. The results of both perceptual spectral subtraction methods performed significantly better than the PLP processing. It is also observed that the proposed speech enhancement performed better than the state of the art spectral subtraction using optimized smoothing and minimum noise statistics.

In speech enhancement techniques utilizing psychoacoustics, the usual procedure is that any noise components that are below the masked threshold will not be detectable by the human listener and so perceptually are not important components to suppress. This is particularly true when the enhanced speech is tested on human subjects. Additionally, we show that, for ASR applications, further improvements in recognition performance may be obtained by augmenting the masking of the noise by additional spectral subtraction in the masked region. The strategy is to remove the masked noise from the ASR system, similar to the masking effect in the human auditory system. This has the effect of improving the SNR without degradation of speech intelligibility. We also tried the subtraction of the noise that lies above the masking threshold, with poor results. These results indicate that in speech recognition applications, spectral subtraction utilizing psychoacoustics may be used for improved recognition performance in noisy conditions, as evaluated by the method of SNR improvement.

156

Chapter 8

Major findings and future research

8.1 Major findings

In this book, we have investigated the role of perceptual features in automatic speech recognition in clean and noisy environments. Our primary objective has been to identify the perceptual features and auditory processing strategies which are relevant for the enhancement of speech recognition performance. In Chapter 3, we investigated the temporal processing of speech using a zero-crossing algorithm and evaluated the performance of auditory adaptation as observed in the peripheral auditory system. The first major finding is that the rapid adaptation could be implemented in temporal processing of speech, not otherwise possible in spectral domain implementation. When applied to ASR, temporal synaptic adaptation demonstrates marginal improvements in white noise at high SNR and in factory noise at low SNR conditions.

Two-tone suppression is implemented by temporal companding utilizing a zero-crossing algorithm for feature extraction. Substantial performance gain can be obtained by the spectral contrast enhancement property of the two-tone suppression, particularly in non-Gaussian real-world noise.

Perceptual features show a high degree of bias on frequency content in the speech articulation, the recognition performance depending on the articulation type. Similar bias is demonstrated in performance in noise conditions, depending on the type of noise and the frequency content therein. A second major finding is that a comparison of noise performance with synaptic adaptation and two-tone suppression indicates that with a zero-crossing algorithm for feature extraction, synaptic adaptation demonstrates greater immunity to white Gaussian noise than the two-tone suppression, while two-tone suppression demonstrates greater immunity to the non-Gaussian, real-world noise than the synaptic adaptation.

In Chapter 5, frequency dependence of compression characteristics in ASR parameterization is investigated utilizing the psychophysical input/output perception curves. Motivated by the mammalian auditory system, a frequency dependent asymmetric compression technique, that is, higher compression in the higher frequency regions than the lower frequency regions, is developed. It is demonstrated that the proposed parametric method may increase compression for improved recognition performance and audibility without degrading the spectral contrasts, and is a simpler alternative to the companding scheme without many of the complex computations involved with it.

Although much has been studied of the role of the peripheral auditory system, only a few studies have been conducted on the role of higher auditory system in automatic speech recognition. In Chapter 6, a novel feature extraction method for ASR front-end is proposed based on the processing strategy in the cochlear nucleus. Features were extracted by the segregated and the parallel processing of three subsections corresponding to the AVCN, DCN and PVCN of the cochlear nucleus, each of which is assigned a particular task of detecting and enhancing a particular speech segment. The main finding is that separating the synchrony detection from synaptic processing, which is consistent with the morphological processing in the CN, may improve ASR performance. It is further observed that a high time constant used in feature extraction from the average discharge rate due to the synaptic processing may improve system performance in clean and white noise conditions.

In Chapter 7, a time-frequency spectral subtraction method based on psychoacoustic properties of human perception is developed for evaluation in an ASR front-end by the method of signal-to-noise ratio improvement. An estimate of the noise and a noise masking threshold is determined from the noisy speech utilizing a detailed psychoacoustic transformation occurring in the peripheral auditory system. It is observed that both synaptic adaptation and the simultaneous masking raise the masking threshold, with the simultaneous masking having greater effects than the synaptic adaptation. It is determined that spectral subtraction of the total perceptual noise from the perceptual noisy speech spectrum improves recognition performance, probably by optimizing in some sense the subtraction algorithm eg., by reducing the residual noise. A significant finding is that the performance may be further improved by utilizing the noise masking threshold, that is, if the masking of noise by the tonal components is augmented by additional spectral subtraction in the masked regions below the masking threshold.

158

8.2 Future research

The following issues may be addressed in any future research in this field.

Besides objective listening tests, ASR may be used to measure the effects of psychoacoustics in speech processing and enhancement. However, measuring performances of auditory models is not standardized yet. A HMM standard baseline for auditory processing with perceptual features for speech recognition was defined in Table 2.4 which may be used as a basis for comparison of auditory models implemented as ASR front-ends. Other objective measures may be developed. Besides statistical methods like HMM other methods, such as artificial neural networks may be utilized for this purpose.

It was observed that in the ZCPA, the application of perceptual features such as synaptic adaptation and two-tone suppression degraded the performance over the baseline performance in white noise when the SNR was very low. The reason for this degradation was explained in Sec 3.3. One way the performance in white noise may be improved is by adjusting the FIR filter order for a desired filter response. For our case, it was chosen empirically and set for optimum recognition performance for a given analysis window over which the zero-crossing intervals are collected, but it probably depends on other performance parameters such as the number of filters, the filter bandwidth and the type of noise. For example, in the minimum variance method of spectrum estimation of a stochastic process, the noise power spectrum estimate at the filter output decreases as the filter order increases [92]. Further research may be undertaken in this direction.

Frequency dependent asymmetric compression may be used to improve ASR performance. Chapter 5 describes two such methods. One of the main limitations of this method is that the equal loudness coefficients depend on the filterbank implementation. Although an analytic expression for its determination have been proposed, this is applicable only to critical band analysis using the Bark scale. Another limitation of the method is that the frequency over which the asymmetric compression is applied is arbitrarily determined. Moreover, the relationship between the compression coefficient and the compression angle is nonlinear. An optimization of compression coefficient with respect to the compression angle may further improve performance. Dynamic asymmetric compression utilizing AGC may also be utilized in auditory models for improved speech intelligibility, particularly for hearing aids and cochlear implants. These may be part of any future research in this field.

It was observed that the mean discharge rate corresponding to the synapse processing was very susceptible to white noise. The ability of the chopper neurons in the PVCN to fire consistently even in presence of severe noise may be one reason for immunity to noise by the human auditory system. In this research, the mean discharge rate features were extracted by an envelope detection process, where a higher time constant helps to reduce the noise variance. Other methods of MR feature extraction may be explored which are less susceptible to noise.

In chapter 7, the experiment exemplifies the differences of psychoacoustics as evaluated on human subjects and by an ASR system, whereby it was determined that the ASR spectral subtraction in the masked regions is often helpful. The proposed method of spectral subtraction may also be evaluated by objective and subjective listening tests. Finally, in masking and short-term synaptic adaptation, effects of the previous frame was not considered. This effect may be included in future research.

8.3 Generalized conclusions

This research has demonstrated that psychoacoustics of human speech perception may be applied to improve automatic speech recognition performances in clean and noisy conditions. Auditory adaptation shows performance improvements in clean conditions and in additive white noise. Two tone suppression performs better in clean conditions and in real-world noise. Compression may be used in ASR parameterizations to improve performance in all signal conditions and additive noise types, both gaussian and non-Gaussian. Speech recognition may also benefit from spectral subtraction methods derived from human masking effects. Specific auditory features demonstrate widely varying frequency bias - some property enhances low frequency segments, other the high frequency segments. Therefore better results are obtained when several perceptual features are combined together in a differential processing scheme, as observed in the higher auditory system.

Processing perceptual features is facilitated by temporal processing utilizing higher time constants and wider window lengths. Temporal processing also assists in capturing dynamic and transient speech segments associated with perceptual features. In speech recognition using short-term methods, the smaller windows used may not be suitable for processing perceptual features. Therefore, other recognition procedures based on higher order discriminative methods and decision surfaces, including ANN methods should also be explored. However, short-term methods appear to perform better in noisy conditions. The main drawback of perceptual processing is that these are computationally expensive. The experiments have demonstrated that simple FIR filters simplify computation and may be used as cochlear filters for

ASR applications.

The benefits demonstrated in this book strongly justify further research in the field of application of perceptual features for automatic speech recognition. Additionally, many of the results may be applied to speech processing, particularly in hearing aids and cochlear implants for improved intelligibility.

Appendix A

Statistical significance test for speech recognition algorithms

(The following derivations are extracts from Gillick and Cox [102]).

Let A_1 and A_2 be two algorithms which are presented with a sequence of labeled isolated utterances of syllables or words $u_i = u_1, u_2, \ldots, u_n$ for recognition by HMM tested on the same data set, and p_1 and p_2 be the true (but unknown) error-rates, respectively, of the two algorithms. It is usually required to determine whether the difference in error-rates between two algorithms tested on the same data set is statistically significant or not. In other words, the problem is equivalent to determining whether there is sufficient evidence to conclude that either $p_1 > p_2$, $p_1 = p_2$ or $p_1 < p_2$.

This problem may be addressed by a statistical approach. If we define the random variable X_i^j such that $X_i^j = 0$ when A_j labels u_i correctly and $X_i^j = 1$ when A_j labels u_i incorrectly, it is reasonable to assume that $S^j = \Sigma_{i=1}^n X_i^j$ follows a binomial distribution $B(n, p_j)$ as long as the errors are independent events, such as in isolated word recognition [102].

The maximum likelihood (ML) estimate of p_j is $\hat{p}_j$

$$\hat{p}_j = S^j / n \tag{1}$$

$\hat{p}_j$ will have a variance σ_j^2 given by

$$\sigma_j^2 = \frac{p_j(1 - p_j)}{n} \tag{2}$$

The null hypothesis to be tested is that there is no difference in the error rates p_1 and p_2, that is

$$\mathbf{H_0} : \qquad \mathbf{p_1 = p_2 = p}; \tag{3}$$

which is equivalent to the hypothesis that $d = p_1 - p_2 = 0$. Under $\mathbf{H_0}$, the ML estimate of d is $\hat{d} = \hat{p}_1 - \hat{p}_2$, with associated variance σ_d^2

$$\sigma_d^2 = Var(\hat{p}_1 - \hat{p}_2) \tag{4}$$

If we can assume that $\hat{p}_1$ and $\hat{p}_2$ are independent events, Eqn. (4) can be written as

$$\sigma_d^2 = Var(\hat{p}_1) + Var(\hat{p}_2) = \sigma_1^2 + \sigma_2^2 \tag{5}$$

σ_d^2 can be estimated by

$$\hat{\sigma}_d^2 = \frac{2\hat{p}(1-\hat{p})}{n} \qquad \text{if } \mathbf{H_0} \text{ is true} \qquad (6)$$

where the ML estimate of p is $\hat{p}=(\hat{p}_1 + \hat{p}_2)/2$. Then if n is large enough and $\mathbf{H_0}$ is true, the distribution of the statistic

$$W = \frac{(\hat{p}_1 - \hat{p}_2)}{\sqrt{\left(\frac{2\hat{p}(1-\hat{p})}{n}\right)}} \qquad (7)$$

tends to a normal distribution with zero mean and unit variance $\aleph(0,1)$. To test the null hypothesis, the expression $P=2\mathrm{Pr}(Z \geq |w|)$ is computed, where Z is a random variable with normal distribution and w is the realizable value of W. If $P < \alpha$ for a chosen significance level α small enough ($\alpha < 0.05$), then $\mathbf{H_0}$ is rejected, which means the difference in error rates are statistically significant, and not by chance.

The above analysis is valid if we assume that $\hat{p}_1$ and $\hat{p}_2$ are independent events. This may not be true when A_1 and A_2 are tested on the same dataset by algorithms which are similar, in which case there may be many errors in common. This problem may be approached by the McNemar's test in which the joint performance of the two algorithms can be stated in a 2×2 matrix form as

$$A_2$$

		Correct	Incorrect
A_1	Correct	N_{00}	N_{01}
	Incorrect	N_{10}	N_{11}

where

$N_{00} = $ No of utterances which both A_1 and A_2 classifies correctly

$N_{01} = $ No of utterances which A_1 classifies correctly, A_2 classifies incorrectly

$N_{10} = $ No of utterances which A_1 classifies incorrectly, A_2 classifies correctly

$N_{11} = $ No of utterances which both A_1 and A_2 classifies incorrectly.

It may be noted that, $n=N_{00}+N_{01}+N_{10}+N_{11}$. By analogy with N_{ij}'s, we may define q_{ij}'s such that $q_{00}=\mathrm{Pr}$(both A_1 and A_2 classifies correctly), etc. Hence, $E\{N_{ij}\} = nq_{ij}$. The parameter $q=q_{10}/(q_{01} + q_{10})$ may be defined as the conditional probability that A_1 will make an error on an utterance given that only one of the two algorithms makes an error. To test $\mathbf{H_0}$, therefore, it is only necessary to examine the utterances on which only one of the algorithms made an error.

Thus for the number of utterances $K=N_{10}+N_{01}$ on which only one algorithm made an error, then for the observed $K=k$, N_{10} has a $B(k,q)$ distribution. Moreover, under $\mathbf{H_0}$, N_{10} has a $B(k,\frac{1}{2})$ distribution. The null hypothesis is thus tested by applying a two-tailed test (as stated in the first part above) to the observation of a random variable M drawn frcm $B(k,1/2)$ distribution.

$$
\begin{align}
P &= 2\Pr(n_{10} \leq M \leq k) \qquad \text{when } n_{10} > k/2 \tag{8}\\
&= 2\Pr(0 \leq M \leq n_{10}) \qquad \text{when } n_{10} < k/2 \tag{9}\\
&= 1.0 \qquad\qquad\qquad\quad \text{when } n_{10} = k/2. \tag{10}
\end{align}
$$

The probabilities can also be directly calculated from

$$
\begin{align}
P &= 2\sum_{m=n_{10}}^{k} \binom{k}{m} \left(\tfrac{1}{2}\right)^{k} \quad \text{when} \quad n_{10} > k/2 \tag{11}\\
&= 2\sum_{m=0}^{n_{10}} \binom{k}{m} \left(\tfrac{1}{2}\right)^{k} \quad \text{when} \quad n_{10} < k/2. \tag{12}
\end{align}
$$

Then as stated above, $\mathbf{H}_0$ is rejected when P is less than some significance level α (usually less than 0.05).

The McNemar's test requires the errors made by an algorithm to be independent events. This is more appropriate for isolated word algorithms which does not make use of any context, such as isolated digits. However, in the case of isolated words that use a language model, the errors will no longer be independent. In the case of continuous words, if each spoken phrase is reasonably short (i.e., a few words, such as in connected digits), it can be considered to be as an entity which is either recognized correctly or incorrectly and are reasonably independent. For such cases, the McNemar's test can also be applied [102].

Appendix B

Analysis of two-tone suppression by companding

Referring to Fig. 4.1, let the input x_0 be a continuous-time signal of the form

$$x_0 = a_1 \sin(\omega_1 t) + a_2 \sin(\omega_2 t + \varphi_0). \tag{13}$$

If the gain (f)and phase (φ) of the pre-filter F at the frequencies ω_1 and ω_2 are given as

$$
\begin{aligned}
f_1 &= |F(j\omega_1)|, \quad f_2 = |F(j\omega_2)| \\
\varphi_1 &= \angle(F(j\omega_1)), \quad \varphi_2 = \angle(F(j\omega_2))
\end{aligned} \tag{14}
$$

then

$$x_1 = f_1 a_1 \sin(w_1 t + \varphi_1) + f_2 a_2 \sin(w_2 t + \varphi_0 + \varphi_2). \tag{15}$$

If it is assumed that the peak detection at the envelop detector is nearly ideal and that the frequency ratio w_1/w_2 is not a small rational number, then the envelop of x_1 at the output of the envelop detector may be approximated by [105]

$$x_{1e} = f_1 a_1 + f_2 a_2. \tag{16}$$

At the output of the compression block,

$$x_2 = x_1 x_{1e}^{(n_1 - 1)}. \tag{17}$$

For the post-filter G, if the gain and the phase at the frequencies ω_1 and ω_2 are expressed as

$$
\begin{aligned}
g_1 &= |G(j\omega_1)|, \quad g_2 = |G(j\omega_2)| \\
\vartheta_1 &= \angle(G(j\omega_1)), \quad \vartheta_2 = \angle(G(j\omega_2))
\end{aligned} \tag{18}
$$

then

$$x_3 = [g_1 f_1 a_1 \sin(w_1 t + \varphi_1 + \vartheta_1) + g_2 f_2 a_2 \sin(w_2 t + \varphi_0 + \varphi_2 + \vartheta_2)] x_{1e}^{(n_1 - 1)}. \tag{19}$$

At the output of the post filter envelop detector, the envelop of x_3, denoted by x_{3e} may be approximated by

$$x_{3e} = (g_1 f_1 a_1 + g_2 f_2 a_2) x_{1e}^{(n_1-1)}. \tag{20}$$

Therefore, the output y_0 of the companding stage may be expressed as given in [105]

$$
\begin{aligned}
y_0 &= x_3 x_{3e}^{((n_2-n_1)/n_1)} \\
&= [g_1 f_1 a_1 \sin(w_1 t + \varphi_1 + \vartheta_1) \\
&\quad + g_2 f_2 a_2 \sin(w_2 t + \varphi_0 + \varphi_2 + \vartheta_2)] x_{1e}^{(n_1-1)} \\
&\quad \cdot \left((g_1 f_1 a_1 + g_2 f_2 a_2) x_{1e}^{(n_1-1)} \right)^{((n_2-n_1)/n_1)} \\
&= [g_1 f_1 a_1 \sin(w_1 t + \varphi_1 + \vartheta_1) \\
&\quad + g_2 f_2 a_2 \sin(w_2 t + \varphi_0 + \varphi_2 + \vartheta_2) \\
&\quad \cdot (g_1 f_1 a_1 + g_2 f_2 a_2)^{((n_2-n_1)/n_1)} x_{1e}^{n_2((n_1-1)/n_1)}.
\end{aligned}
\tag{21}
$$

If the pre and post filters have a resonance frequency of w_1, then $g_1 = f_1 = 1$. Moreover, if (G is sharply tuned and w_2 is distant from w_1), then $g_2 = 0$. For such cases, the output y_0 may be simplified as

$$
\begin{aligned}
y_0 &= \left[a_1^{n_2/n_1} (a_1 + f_2 a_2)^{n_2(n_1-1)/n_1} \right] \sin(\omega_1 t + \varphi_1 + \vartheta_1). \\
&= \left[a_1 \left(\frac{a_1 + f_2 a_2}{a_1} \right)^{((n_1-1)/n_1)} \right]^{n_2} \sin(\omega_1 t + \varphi_1 + \vartheta_1).
\end{aligned}
\tag{22}
$$

It is seen from Eqn. (22) that the presence of second tone with amplitude a_2 suppresses the tone with amplitude a_1. If only a single tone is present such that $a_2 = 0$, then

$$y_0 = \sin(w_1 + \varphi_1 + \vartheta_1) a_1^{n_2}. \tag{23}$$

If $n_2 = 1$, then the output has the amplitude of a_1 only, that is, no compression or expansion takes place.

166

Appendix C

C.1 Normalized equal-loudness weighting coefficients (L_m) for 40 channel mel-frequencies.

Mel frequency, CF (Hz)	Loudness level in phons									
	10-19	20-29	30-39	40-49	50-59	60-69	70-79	80-89	90-99	100-109
10	1.000	1.000	1.000	1.000	1.000	1.000	1.000	1.000	1.000	1.000
199	0.220	0.325	0.395	0.450	0.526	0.600	0.638	0.705	0.725	0.757
266	0.194	0.300	0.348	0.428	0.494	0.580	0.629	0.705	0.725	0.757
333	0.168	0.287	0.337	0.417	0.494	0.560	0.611	0.678	0.720	0.757
399	0.142	0.262	0.325	0.406	0.484	0.560	0.601	0.669	0.716	0.757
466	0.129	0.250	0.325	0.395	0.484	0.560	0.592	0.678	0.716	0.757
533	0.129	0.250	0.325	0.395	0.484	0.560	0.592	0.678	0.725	0.765
599	0.129	0.250	0.325	0.384	0.494	0.570	0.601	0.678	0.725	0.765
666	0.129	0.250	0.325	0.384	0.505	0.570	0.611	0.687	0.733	0.765
733	0.129	0.250	0.337	0.395	0.505	0.580	0.620	0.696	0.733	0.773
799	0.129	0.250	0.337	0.406	0.515	0.580	0.629	0.696	0.741	0.773
866	0.129	0.250	0.337	0.417	0.515	0.590	0.638	0.705	0.741	0.781
933	0.129	0.250	0.337	0.428	0.526	0.600	0.648	0.714	0.751	0.781
999	0.129	0.250	0.337	0.439	0.526	0.600	0.648	0.714	0.751	0.781
1070	0.129	0.250	0.348	0.439	0.526	0.600	0.648	0.714	0.750	0.781
1147	0.129	0.250	0.348	0.439	0.494	0.600	0.648	0.705	0.750	0.781
1228	0.129	0.250	0.348	0.428	0.494	0.590	0.648	0.705	0.750	0.781
1316	0.129	0.250	0.348	0.428	0.494	0.590	0.638	0.705	0.741	0.773
1409	0.129	0.250	0.337	0.428	0.505	0.590	0.638	0.696	0.741	0.765
1510	0.129	0.250	0.337	0.417	0.505	0.580	0.629	0.696	0.733	0.757
1617	0.129	0.250	0.325	0.417	0.505	0.580	0.629	0.705	0.725	0.757
1732	0.129	0.237	0.325	0.417	0.494	0.580	0.620	0.687	0.716	0.750
1856	0.129	0.225	0.313	0.406	0.494	0.570	0.620	0.687	0.708	0.742
1988	0.116	0.212	0.313	0.406	0.494	0.570	0.611	0.687	0.708	0.734
2129	0.103	0.200	0.313	0.395	0.494	0.570	0.611	0.678	0.700	0.734
2281	0.090	0.187	0.302	0.384	0.484	0.570	0.601	0.669	0.691	0.726
2443	0.070	0.175	0.290	0.393	0.473	0.550	0.592	0.660	0.683	0.718
2617	0.060	0.162	0.279	0.362	0.460	0.530	0.583	0.651	0.683	0.710
2803	0.052	0.150	0.267	0.351	0.450	0.520	0.574	0.642	0.675	0.703
3003	0.038	0.150	0.255	0.351	0.442	0.520	0.564	0.625	0.675	0.695
3217	0.026	0.150	0.255	0.351	0.431	0.520	0.564	0.625	0.675	0.675
3446	0.026	0.150	0.255	0.340	0.431	0.520	0.564	0.625	0.660	0.687
3691	0.026	0.150	0.255	0.351	0.421	0.510	0.555	0.625	0.660	0.687
3954	0.038	0.151	0.255	0.351	0.421	0.510	0.555	0.625	0.660	0.687
4235	0.052	0.175	0.255	0.351	0.431	0.510	0.555	0.625	0.675	0.687
4537	0.060	0.200	0.267	0.362	0.442	0.510	0.564	0.634	0.675	0.695
4860	0.070	0.212	0.279	0.279	0.362	0.450	0.520	0.574	0.634	0.683
5205	0.070	0.225	0.279	0.373	0.450	0.520	0.574	0.642	0.683	0.703
5576	0.103	0.237	0.302	0.417	0.484	0.550	0.648	0.669	0.691	0.726
5973	0.129	0.250	0.325	0.439	0.515	0.580	0.648	0.714	0.708	0.750
6398	0.155	0.262	0.348	0.461	0.547	0.600	0.740	0.714	0.733	0.757

C.2 Critical Bands and the Bark frequency scale [41].

Center frequency Hz	Critical bandwidth Hz	Edge frequency Hz	CB-rate Bark
50	80	100	1
150	100	200	2
250	100	300	3
350	100	400	4
450	110	510	5
570	120	630	6
700	140	770	7
840	150	920	8
1000	160	1080	9
1170	190	1270	10
1370	210	1480	11
1600	240	1720	12
1850	280	2000	13
2150	320	2320	14
2500	380	2700	15
2900	450	3150	16
3400	550	3700	17
4000	700	4400	18
4800	900	5300	19
5800	1100	6400	20
7000	1300	7700	21
8500	1800	9500	22
10500	2500	12000	23
13500	3500	15500	24

Bibliography

[1] L. Deng and D. O'Shaughnessy, *Speech Processing - A Dynamic and Optimization-Oriented Approach*, Marcel Dekker, Inc., New York, 2003.

[2] B. Gold and N. Morgan, *Speech and Audio Signal Processing: Processing and Perception of Speech and Music*, John Wiley and Sons, Inc., New York, 2000.

[3] W. Koenig, H. Dunn, and L. Lacy, "The sound spectrograph," *J. Acoust. Soc. of Am.*, vol. 18, no. 1, pp. 19-49, 1946.

[4] J. D. Markel and A. H. Gray, *Linear Prediction of Speech*, Springer-Verlag, Berlin, 1976.

[5] B. P. Bogert, M. J. Healy, and J. W. Tukey, "The quefrency analysis of time series for echoes: cepstrum, pseudo-autocovariance, cross-cepstrum, and shape tracking," M. Rosenblatt, editor, *Time Series Analysis*, pages 209-243, John Wiley, New York, 1963.

[6] A. M. Abdelatty, J. V. Spiegel, P. Mueller, G. Haentjens, and J. Berman, "An Acoustic-phonetic feature-based system for automatic phoneme recognition in continuous speech," *IEEE Int. Conf. on Acoust. Speech, Signal Processing* (ICASSP 1999).

[7] D. S. Kim, S. Y. Lee, and R. M. Kil, "Auditory processing of speech signals for robust speech recognition in real world noisy environments," *IEEE Trans. Speech Audio Processing*, vol. 7, no. 1, pp. 55-69, Jan. 1999.

[8] B. Gajić and K. K. Paliwal, "Robust speech recognition in noisy environments based on subband spectral centroid histograms," *IEEE Trans. Audio, Speech, Lang. Processing*, vol. 14, no. 2, pp. 600-608, Mar. 2006.

[9] S. Seneff, "A joint synchrony/mean-rate model of auditory processing," *J. Phonetics*, vol. 16, pp. 55-76, 1988.

169

[10] O. Ghitza, "Temporal nonplace information in the auditory-nerve firing pattern as a front-end for speech recognition in a noisy environment," *J. Phonetics*, vol. 16, no. 1, pp-76, Jan. 1988.

[11] R. F. Lyon and C. Mead, "An analog electronic cochlea," *IEEE Trans. Acoust., Speech, Signal Processing*, vol. 36, pp. 1119-1134, Jul. 1988.

[12] X. Zhang, M. G. Heinz, I. C. Bruce, and L. H. Carney, "A phenomenological model for the responses of auditory-nerve fibres: I. Nonlinear tuning with compression and suppression," *J. Acoust. Soc. Am.*, vol. 109. no. 2, Feb. 2001.

[13] T. Baer, B. C. J. Moore, and S. Gatehouse, "Spectral contrast enhancement of speech in noise for listeners with sensorineural hearing impairment: Effects on intelligibility, quality, and response times," *J. Rehabil. Res. Dev.*, vol. 30, no. 1, pp. 49-72, 1993.

[14] D. B. Grayden, A. N. Burkitt, O. P. Kenny, J. N. Clarey, A. G. Paolini, and G. M. Clark, "A cochlear implant speech processing strategy based on an auditory model," *Int. Conf. on Intelligent Sensors, Sensor Networks and Information Processing* (ISSNIP 2004), pp. 491-496, Melbourne.

[15] M. J. Hunt and C. Lefebvre, "Speech recognition using a cochlear model," *IEEE Int. Conf. Acoust. Speech, Signal Processing* (ICASSP 1986), Tokyo.

[16] J. R. Cohen, "Application of an auditory model to speech recognition," *J. Acoust. Soc. Am.*, vol. 85, no. 6, pp. 2623-2629, Jun. 1989.

[17] E. Zwicker, E. Terhardt, and E. Poulus, "Automatic speech recognition using psychoacoustic models," *J. Acoust. Soc. Am.*, vol. 65. No. 2, Feb. 1979.

[18] C. L. Searle, J. Jacobson, and S. G. Rayment, "Stop consonant discrimination based on human audition," *J. Acoust. Soc. Am.*, vol. 65, pp. 799-809, 1979.

[19] O. Ghitza, "Auditory models and human performance in tasks related to speech coding and speech recognition," *IEEE Trans. Speech Audio Processing*, vol. 2, no. 1, pp. 115-132, Jan. 1994.

[20] M. Bloomberg, R. Carlson, K. Elenius, and B. Granstrom, "Auditory models and isolated word recognition," *Q Prog. Stat. Rep.*, Speech Transmiss. Lab. (Royal Institute of Technology, Stockholm), 1-15, 1984.

[21] C. R. Jankowski Jr., H. H. Vo, and R. P. Lippman, "A comparison of signal processing front ends for automatic word recognition," *IEEE Trans. Speech Audio Processing*, vol. 3. pp. 286-293, Jul. 1995.

[22] J. Tchorz and B. Kollmeier, "A model of auditory perception as front end for automatic speech recognition," *J. Acoust. Soc. Am.*, vol. 106(4), Pt. 1, Oct. 1999.

[23] S. Sandhu and O. Ghitza, "A comparative study of mel cepstra and EIH for phone classification under adverse conditions," *IEEE Int. Conf. Acoust. Speech, Signal Processing* (ICASSP 1995).

[24] B. Strope and A. Alwan, "A model of dynamic auditory perception and its application to robust word recognition," *IEEE Trans. Speech Audio Processing*, vol. 95. no. 5, pp. 451-464, Sep. 1997.

[25] M. B. Sachs, H. F. Voigt, and E. D. Young, "Auditory nerve representation of vowels in background noise," *J. Neurophysiol.*, vol. 50, no. 1, Jul. 1983.

[26] P. Loughlin, D Groutage, and R. Rohrbaugh, "Time-frequency analysis of acoustic transients," *IEEE Int. Conf. Acoust. Speech, Signal Processing* (ICASSP 1997).

[27] M. Holmberg, D. Gelbart, and W. Hemmert, "Automatic Speech Recognition with an adaptation model motivated by auditory processing," *IEEE Trans. Audio, Speech, Lang. Processing*, vol. 14, no. 1, pp. 44-49, Jan. 2006.

[28] J. W. Pitton, K. Wang, and B. Juang, "Time-Frequency analysis and auditory modeling for automatic recognition of speech," *Proc. of the IEEE*, vol. 84, no. 9, Sep. 1996.

[29] B. Kedem, "Spectral analysis and discrimination by zero-crossings," *Proc. of the IEEE*, vol. 74, no. 11, pp. 1477-1492, Nov. 1986.

[30] P. F. Castalez and R. J. Niederjohn, "A comparative study of the use of zero-crossing analysis methods for vowel recognition," *IEEE Int. Conf. Acoust. Speech and Signal Processing* (ICASSP 1976).

[31] L. Westerman and R.L. Smith, "Rapid and short-term adaptation in auditory nerve responses," *The Hearing Review*, 15, pp. 249-260, 1984.

[32] R. D. Patterson, "Time interval information in the auditory representation of speech sounds," *J. Acoust. Soc. of Am.*, vol. 105, no. 2, Feb. 1999.

[33] H. Hermansky, "Speech Beyond 10 ms (Temporal Filtering in Feature Domain)," *Proc. of the International Workshop on Human Interface Technology*, Aizu, Japan, Sep. 1994.

[34] M. B. Sachs and E. D. Young, "Encoding of steady state vowels in the auditory-nerve: Representation in terms of discharge rate," *J. Acoust. Soc. Am.*, vol. 66, pp. 470-479, 1979.

[35] J. O. Pickles, *An Introduction to the Physiology of Hearing*, Academic, London, 1982.

[36] N. Y. S Kiang and W. T. Peake, "Physics and physiology of hearing," *Stevens Handbook of Experimental Psychology*, 2nd ed., New York: Wiley, 1988, pp. 277-326.

[37] W. S. Rhode and S. Greenberg, *Physiology of the Cochlear Nuclei*, Springer handbook of auditory research, vol. 2, Chapter 3, Springer-Verlag, New York, 1992, pp. 94-152.

[38] B. Xiang, X. Wu, Z. Liu, and H. Chi, "Auditory model based feature extraction and its application to speaker identification," *IEEE Int. Conf. Acoust. Speech and Signal Processing* (ICASSP 1999).

[39] G. von Békésy, *Experiments in Hearing*, McGraw-Hill, New York, 1960.

[40] D. Greenwood, "A cochlear frequency-position function for several species-29 years later," *J. Acoust. Soc. Am.*, vol. 87, no. 6, pp. 2592-2605, 1990.

[41] E. Zwicker, G. Flottorp, and S. S. Stevens, "Critical bandwidth in loudness summation," *J. Acoust. Soc. Am.*, 29, pp. 548-557, 1957.

[42] B. C. J. Moore and B. R. Glasberg, "Suggested formulae for calculating auditory filter bandwidths and excitation patterns," *J. Acoust. Soc. Am.*, vol. 74, no. 3, Sep. 1983.

[43] A. Hodgkin and A. Huxley, "A quantative description of membrane current and its application to conduction and excitation in nerves," *J. Physiology*, vol. 117, pp. 500-544, 1952.

[44] L. Deng and D. Geisler, "A composite auditory model for processing speech sounds," *J. Acoust. Soc. Am.*, vol. 82, no. 6, pp. 2001-2012, 1987.

[45] R. Meddis, "Simulation of mechanical to neural transduction in the auditory receptor," *J. Acoust. Soc. Am.*, vol. 79(3), Mar. 1988.

[46] M. B. Sachs, C. C. Blackburn, and E. D. Young, "Rate-place and temporal-place representations of vowels in the auditory nerve and anteroventral cochlear nucleus," *J. Phonet*, vol. 16, pp. 37-53, 1988.

[47] B. Delgutte and N. Y. S. Kiang, "Speech coding in the auditory nerve: I," *J. Acoust. Soc. Am.*, vol. 75, pp. 866-878, 1984.

[48] L. Kinsler and A. Frey, *Fundamentals of Acoustics*, John Wiley and Sons, Inc., New York, 1962.

[49] B. C. J. Moore, *An Introduction to the Psychology oh Hearing*, 3rd. ed., Academic Press, New York/London, 1989.

[50] H. Fletcher and W. J. Munson, "Loudness, its definition, measurement and calculation," *J. Acoust. Soc. Am.*, vol. 5, pp. 82-108, 1933.

[51] M. Vlaming, "Concave curvilinear WDRC: optimizing the shape of compression," *The Hearing Review*, Sep. 2000.

[52] M. A. Ruggero, N. C. Rich, A. Recio, S. S. Narayan, and L. Robles, "Basilar-membrane responses to tones at the base of the chinchilla cochlea," *J. Acoust. Soc. Am.*, 101, 2151-2163, 1997.

[53] X. Zhang, M. G. Heinz, and L. H. Carney, "Nonlinear compression in an auditory-nerve model," *Proc. of The First Joint BMES/EMBS Conf. Serving Humanity, Advancing Technology*, pp. 13-16, Atlanta, Oct. 1999.

[54] R. Smith and J. J. Zwislocki, "Short-term adaptation and incremental responses of single auditory receptor," *Biological Cybernetics*, 17, pp. 169-182, 1975.

[55] A. J. Oxenham, "Forward masking: adaptation or integration?" *J. Acoust. Soc. Am.*, vol. 109, pp. 732-741, 2001.

[56] D. M. Harris and P. Dallos, "Forward masking of auditory nerve fiber responses," *J. Neural Physiol.*, vol. 42, pp. 1083-1107, Jul. 1979.

[57] M. R. Schroeder, B. S. Atal, and J. L. Hall, "Optimizing digital speech coders by exploiting masking properties of the human ear," *J. Acoust. Soc. Am.*, vol. 66, no. 6, Dec. 1979.

[58] B. Delgutte, "Two-tone rate suppression in auditory nerve fibers: Dependence on suppressor frequency and level," *Hear. Res.*, vol. 49, pp. 225-246, 1990.

[59] W. S. Rhode, "Some observations on cochlear mechanics," *J. Acoust. Soc. Am.*, vol. 64, no. 1, pp. 158-176, 1978.

[60] R. M. Arthur, R. R. Pfeiffer, and N. Suga, "Properties of two-tone inhibition in primary auditory neurons," *J. Physiol.*, vol. 212, pp. 593-609, 1971.

[61] K. A. Davis and H. F. Voigt, "Neural modeling of the DCN: cross-correlation analysis using short-duration tone-burst stimuli," *Proceedings of IEEE*, 1991.

[62] T. Houtgast, "Psychophysical evidence for lateral inhibition in hearing," *J. Acoust. Soc. Am.*, vol. 51, no. 6 (part2), pp. 1885-1894, 1972.

[63] F. Ratliff, *"Mach Bands. Quantitative Studies on Neural Networks in the Retina,"* Holden-Day, San Francisco, 1965.

[64] J. Clark and C. Yallop, *"An Introduction to Phonetics and Phonology,"* 2nd Ed. Blackwell, Oxford, UK, 1995.

[65] F. Katamba, *"An Introduction to Phonology,"* Longman, London, 1989.

[66] X. Yang, K. Wang, and S. A. Shamma, "Auditory representations of acoustic signals," *IEEE Trans. Inform. Theory*, vol. 38, pp. 824-839, 1992.

[67] S. B. Davis and P. Mermelstein, "Comparison of parametric representations for monosyllabic word recognition in continuously spoken sentences," *IEEE Trans. Acoust., Speech, Signal Processing*, 28(4), pp. 357-366, Aug. 1980.

[68] S. Young, Gunnar Everman, Thomas Hain, Dan Kershaw, Gareth Moore, Julian Odell, Dave Ollason, Dan Povey, Valtcho Valtchev, and Phil Woodland, *"The HTK Book,"* v 3.2.1, Cambridge University Engineering Department, Dec. 2002.

[69] N. Ahmed, T. Natarajan, and K. R. Rao, "Discrete cosine transform," *IEEE Trans. Comput.*, **L-23**: pp. 90-93, 1974.

[70] H. Hermansky, "Perceptual linear prediction (PLP) analysis of speech," *J. Acoust. Soc. Am.*, vol. 87, pp. 1738-1752, 1990.

[71] H. Hermansky and N. Morgan, "RASTA processing of speech," *IEEE Trans. Speech Audio Processing*, vol. 2, pp. 587-589, Oct. 1994.

[72] S. Furui, "Speaker-independent isolated word recognition using dynamic features of speech spectrum," *IEEE Trans. Acoust., Speech, Signal Processing*, vol. 34, no. 1, pp. 52-59, Feb. 1986.

[73] M. Heinz, S. Colburn, and L. Carney, "Evaluating auditory performance limits: I. One-parameter discrimination using a computational model for auditory nerve," *Neural Computation*, 13, pp. 2273-2316.

[74] D. Duda, and P. Hart, *Pattern Classification and Scene Analysis*, Wiley-Interscience, New York, 1973.

[75] R. Bellman and S. Dreyfus, *Applied Dynamic Programming*, Princeton University Press, Princeton, 1962.

[76] T. Vintsyuk, "Speech discrimination by dynamic programming," *Kibernetika*, 4: 81-88, 1968.

[77] A. Zaknich, *Neural Networks for Intelligent Signal Processing*, World Scientific, Singapore, 2003.

[78] S. Theodoridis and K. Koutroumbas, *Pattern Recognition*, Academic Press, 1999.

[79] L. Deng, "A generalized hidden Markov model with state conditioned trend functions of time for speech signal," *Signal Processing*, vol. 27, pp. 65-78, 1992.

[80] R. Stern and R. A. Cole, Eds., "Survey of the State of the Art in Human Language Technology," *Center for Spoken Language Understanding*, Oregon Graduate Institute, USA, 1996.

[81] C. Wu, Y. Chiu, and H. Lim, "Perceptual speech modeling for noisy speech recognition," *IEEE Int. Conf. Acoust. Speech and Signal Processing* (ICASSP 2002).

[82] P. C. Woodland, D. Pye, and M. J. F. Gales, "Iterative unsupervised adaptation using maximum likelihood linear regression," *Fourth Int. conf. on Spoken Languauge* (ICSLP 1996).

[83] S. Seneff, "Pitch and spectral estimation of speech based on auditory synchrony model," *IEEE Int. Conf. on Acoust. Speech Signal Processing*, San Diego (ICASSP 1984).

[84] L. Bu, "Perceptual speech processing and phonetic feature mapping for robust vowel recognition," *IEEE Trans. Speech Audio Processing*, vol. 8, no. 2, pp. 105-114, Mar. 2000.

[85] M. R. Schroeder and J. L. Hall, "Model for mechanical to neural transduction in the auditory receptor," *J. Acoust. Soc. Am.*, 55, pp. 1055-1060, 1974.

[86] C. J. Sumner, E. A. Lopez-Poveda, L. P. O'Mard, and Ray Meddis, "A revised model of the inner-hair cell and auditory-nerve complex," *J. Acoust. Soc. Am.*, vol. 111. no. 5, May. 2002.

[87] A. M. Abdelatty, J. V. Spiegel, and P. Mueller, "Robust auditory-based speech processing using the average localized synchrony detection," *IEEE Trans. Speech Audio Processing*, vol. 10, no. 5, pp. 279-292, Jul. 2002.

[88] F. Perdigao and L. Sa, "Auditory models as front-ends for speech recognition," *Proc. NATO ASI Computational Hearing*, pp. 179-184, Il Ciocco, Italy, 1998,

[89] A. Spoor, J. J. Eggermont, and D. W. Odenthal, "Comparison of human and animal data concerning adaptation and masking of eighth nerve compound action potential," *Electrocochleography*, Ruber, J, Elberling, C. and Solomon, G., Eds., Baltimore, MD: University Park, pp. 183-198, 1976.

[90] M. Ghulam, T. Fukuda, J. Horikawa and T. Nitta, "Pitch-synchronous ZCPA-based feature extraction with auditory masking," *IEEE Int. Conf. Acoust. Speech and Signal Processing* (ICASSP 2005).

[91] L. H. Koopmans, *The Spectral Analysis of Time Series*, Academic Press, New York, 1974.

[92] M. H. Hayes, *Statistical Digital Signal Processing and Modeling*, John Wiley and Sons, Inc., New Jersey, 1996.

[93] S. Haque, R. Togneri, and A. Zaknich, "A temporal auditory model with adaptation for automatic speech recognition," *IEEE International Conference on Audio, Speech and Signal Processing* (ICASSP 2007).

[94] T. S. Gunawan and E. Ambikairajah, "On the use of simultaneous and temporal masking in noise suppression applications," *Proc. of 11th Int. Conf. Speech Science and Technology* (SST 2006), Auckland.

[95] E. Zwicker and H. Fastl, *Psychoacoustics: Facts and Models*, 2nd ed. New York: Springer-Verlag, 1990.

[96] B. Gajić and K. K. Paliwal, "Robust speech recognition using features based on zero crossings with peak amplitudes," *IEEE Int. Conf. Acoust. Speech, Signal Processing* (ICASSP 2003).

[97] S. Haque, R. Togneri, and A. Zaknich, "Zero-Crossings with adaptation for automatic speech recognition," *The Eleventh Australasian International Conference on Speech Science and Technology* (SST 2006), Auckland.

[98] G. J. McLaclan and K. E. Basford, *Mixture Models: Inference and Applications to Clustering*, Marcel Dekker, Inc., 1988.

[99] G. J. McLaclan, *Discriminant Analysis and Statistical Pattern Recognition*, John Wiley and Sons, New York, 2005.

[100] S. Kullback and R. A. Liebler, "On information and sufficiency," *Annals of Mathematical Statistics*, vol. 22, pp. 79-86, 1951.

[101] C. Lee and D. A. Landgrebe, "Decision boundary feature extraction for neural neteworks," *IEEE trans. Neural Networks*, vol. 8, pp. 75-83, Jan. 1997.

[102] L. Gillick and S. J. Cox, "Some statistical issues in the comparison of speech recognition algorithms," *IEEE Int. Conf. Acoust. Speech, Signal Processing* (ICASSP 1989), Tokyo.

[103] M. A. Ruggero, S. S. Narayan, A. N. Temchin, and A. Recio, "Mechanical bases of frequency tuning and neural excitation at the base of the cochlea," *Proc. Natl. Acad. Sci.*, USA, vol 97, pp. 11744-11750, 2000.

[104] R. L. Miller, B. M. Calhoun, and E. D. Young, "Contrast enhancement improves the representation of e-like vowels in the hearing-impaired auditory nerve," *J. Acoust. Soc. Am.*, vol. 106. no. 5, pp. 2693-708, Nov. 1999.

[105] L. Turicchia and R. Sarpeshkar, "A bio-inspired companding strategy for spectral enhancement," *IEEE Trans. Speech Audio Processing*," vol. 13, no. 2, Mar. 2005.

[106] J. M. Kates, "Two-tone suppression in a cochlear model," *IEEE Trans. Speech Audio Processing*," vol. 3. no. 5, Sep. 1995.

[107] R. Dolby, "An audio noise reduction system," *J. Audio Engg. Soc.*, vol. 15, no. 4, Oct. 1967.

[108] Y. Tsividis, "Externally linear, time-invariant systems and their application to companding signal processors," *IEEE Trans. Circuits Sys. II*, vol. 44, no. 2, pp. 65-85, Feb. 1997.

[109] D. Frey, Y. Tsividis, G. Efthivoulidis, and N. Krishnapura, "Syllabic- Companding log domain filters," *IEEE Trans. Circuits Syst. II*, vol. 48, no. 4, pp. 329-339, Apr. 2001.

[110] B. Moore and B. Glasberg, "A model of loudness perception applied to cochlear hearing loss," *Auditory Neuroscience*, vol. 3, pp. 289-311, 1997.

[111] B. Moore and A. Oxenham, "Psychoacoustic consequences of compression in the peripheral auditory system," *Psychological Review*, 105(1), pp. 108-124, 1998.

[112] S. P. Bacon, R.R. Fay, and A. N. Pooper, *Compression-From Cochlea to Cochlear Implant*, Springer, 2004.

[113] W. S. Rhode and A. Recio, "Study of mechanical motions in the basal region of the chinchilla cochlea," *J. Acoust. Soc. Am.*, vol. 107, pp. 3317-3332, 2000.

[114] C. J. Plack and Vit Drga, "Psychophysical evidence for auditory compression at low charactersitic frequencies," *J. Acoust. Soc. Am.*, vol. 113(3), pp. 1574-1585, Mar. 2003.

[115] D. Puschel, *Prinzipien der zeitlichen Analyse beim Horen*, PhD thesis, University of Gottingen, 1988.

[116] T. Dau, D. Puschel, and A. Kohlrausch, "A quantative model of the "effective" signal processing in the auditory system. I. Model structure," *J. Acoust. Soc. Am.*, 99(6), Jun. 1996.

[117] V. Hohmann and B. Kollmeier, "The effect of multichannel dynamic compression on speech intelligibility," *J. Acoust. Soc. Am.*, vol. 97, no. 2, pp. 1191-1195, 1997.

[118] Matlab, (*Signal Processing Toolbox*), version 7.1.0.183 (R14) Service pack 3, The Mathworks, Inc., Jun. 2005.

[119] C. E. Molnar and R. R. Pfeiffer, "Interpretation of spontaneous spike discharge patterns of neurons in the cochlear nucleus," *Proc. of the IEEE*," vol. 56, no. 6, Jun. 1968.

[120] D. R. Kipke and K. L. Levy, "An application of neural speech processing in the cochlear nucleus to the estimation of fundamental frequency," *Proc. IEEE-EMBC and CMBEC:Theme 4: Signal Processing* pp. 973-974, 1995.

[121] S. A. Shamma, "Speech processing in the auditory system. II. Lateral inhibition and the central processing of speech evoked activity in the auditory nerve," *J. Acoust. Soc. Am.*, vol. 78. no. 5, Nov. 1985.

[122] C. Kim, Y. Chiu, and R. Stern, "Physiologically-motivated synchrony-based processing for robust automatic speech recognition," *Int. Conf. Spoken Lang. Processing* (ICSLP 2006).

[123] A. Guérin, J. Bès, and R. Le Bouquin Jeannès, "A computer model for the temporal envelop encoding in the auditory pathway," *Proc. 25th Annual Int. Conf. of the IEEE EMBS*," Sep. 17-21, 2003, Cancun.

[124] W. M. Siebert, "Frequency discrimination in the auditory system: Place or periodicity mechanism?" *Proc. of the IEEE*, vol. 58, pp. 723-730, 1970.

[125] P. X. Joris, L. H. Carney, P. H. Smith, and T. C. Yin, "Enhancement of neural synchronization in the anteroventral cochlear nucleus. I. Response to tones at the characteristic frequency," *J. Neurophysiol.*, 71, pp. 1022-1036, 1994.

[126] H. Wang, M. Holmberg and W. Hemmert, "Auditory information coding by cochlear nucleus onset neurons," *IEEE Int. Conf. Acoustics, Speech, Signal Processing* (ICASSP 2006).

[127] E. D. Young, I. Nelken, and *et al.*, "Somatosensory effects on neurons in dorsal cochlear nucleus," *J. Neurophysiol.*, 73(2): pp. 743-65, 1995.

[128] E. D. Young and W. E. Brownell, "Responses to tones and noise of single cells in dorsal cochlear nucleus of unanesthetized cats," *J. Neurophysiol.*, vol. 39, pp. 282-300, 1976.

[129] Malcolm Slaney, "*Auditory Toolbox: A Matlab Toolbox for Auditory Modeling Work,*" version 2, Interval Research Corporation, 1994.

[130] K. K. Wójcicki, B. J. Shannon, and K. K. Paliwal, "Spectral subtraction with variance reduced noise spectrum estimate," *Australasian Int. Conf. on Speech Science and Technology* (SST 2006), Auckland.

[131] S. F. Boll, "Suppression of acoustic noise in speech using spectral subtraction," *IEEE Trans. Acoust., Speech, Signal Processing*, vol. 27, pp. 113-120, Apr. 1979.

[132] M. Berouti, R. Schwartz, and L. Makhoul, "Enhancement of speech corrupted by acoustic noise," *IEEE Int. Conf. Acoust. Speech, Signal Processing* (ICASSP 1979), Washington, DC.

[133] R. J. McAulay and M. L. Malpass, "Speech enhancement using a soft decision noise suppression filter," *IEEE Trans. Acoust., Speech, Signal Processing*, vol. ASSP-28, pp. 137-145, Apr. 1980.

[134] Y. Ephraim and D. Malah, "Speech enhancement using a minimum mean-square error short-time spectral amplitude estimator," *IEEE Trans. Acoust., Speech, Signal Processing*, vol. ASSP-32, pp. 1109-1121, Dec. 1984.

[135] P. Lockwood and J. Boudy, "Experiments with a nonlinear spectral subtractor (NSS), hidden Markov models and projection, for robust recognition in cars," *Speech Commun.*, vol. 11, pp. 215-228, Jun. 1992.

[136] J. H. L. Hansen, "Morphological constrained feature enhancement with adaptive cepstral compensation (MCE-ACC) for speech recognition in noise and Lombard effect," *IEEE Trans. Speech Audio Processing*, vol. 2, pp. 598-614, Oct. 1994.

[137] R. Martin, "Noise power spectral density estimation based on optimal smoothing and minimum statistics," *IEEE Transactions on Speech and Audio Processing*, vol. 9. No. 5, Jul. 2001.

[138] A. Spanias, T. Painter, and V. Atti, *Audio Signal Processing and Coding*, John Wiley and Sons, New York, 2006.

[139] H. Gustafsson, S. E. Nordholm, and I. Classen, "Spectral subtraction using reduced delay convolution and adapative averaging," *IEEE Trans. Speech Audio Processing*, 9(8), pp. 798-807, 2001.

[140] L. Lin, E. Ambikairajah, and W. H. Holmes, "Suband noise estimation for speech enhancement using a perceptual Wiener filter," *IEEE Int. Conf. Acoust. Speech, Signal Processing* (ICASSP 2003).

[141] D. E. Tsoukalas, J. N. Mourjopoulos, and G. Kokkinakis, "Speech enhancement based on audible noise suppression," *IEEE Trans. Speech Audio Processing*, 5(6), pp. 497-513, 1997.

[142] K. H. Arehart, J. H. L. Hansen, S. Gallant, and L.Kalstein, "Evaluation of an auditory masked threshold noise suppression algorithm in normal-hearing and hearing-impaired listeners," *Speech Communication*, no. 40, pp. 575-592, 2003.

[143] Yiu-Pong LAI, Man-Chun HUI, Chi-Wah KOK, and Man-Hung SIU, "Speech recognition enhancement by psychoacoustic modeled noise suppression," *IEEE Int. Conf. on Multimedia and Expo* (ICME 2004).

[144] Y. Shao and C.H. Chang, "A generalized perceptual time-frequency subtraction method for speech enhancement," *IEEE Int. Symposium on Circuits and Systems* (ISCAS 2006).

[145] T. S. Gunawan and E. Ambikairajah, "Speech enhancement using temporal masking fractional Bark gammatone filters," *Proc. of the 10th Int. Conf. Speech Science and Technology* (SST 2004), Sydney.

Index

Acoustic modelling, 26, 38

Acoustic pre-processing, 27

Action potential
    See Neural discharge rate

Adaptation time constant, 58, 67

Anteroventral cochlear nucleus (AVCN), 120, 122

Asymmetric compression coefficient, 107

Asymmetric compression in the MFCC, 104

Auditory adaptation
    See Synaptic adaptation

Auditory masking threshold (AMT), 145

Auditory
    masking, 21
    model, 46
    nerve, 11
    system (See Peripheral auditory system)

Autocorrelation function, 31

Automatic gain control, 10

Automatic speech recognition, 1, 36

Average localized synchronized rate (ALSR), 16

Bark frequency scale, 13

Basilar membrane, 12

Baum-Welch re-estimation, 43

Bayes's decision rule, 42

Central auditory system, 119

Cepstral coefficient, 31

Characteristic frequency, 12

Class separability measure, 77
Classifiers
    Bayesian MAP classsifier, 41
    Mahanalobis classifier, 37
    Minimum distance classifier, 37
Cochlea, 7
Cochlear model, 3, 15
Cochlear nucleus, 119
    Fusiform cell, 121
    Octopus cell, 121
    Stellate cell, 121
Companding, 86, 136
    architecture, 86
    with zero-crossings, 88
Compressive nonlinearity, 19
Compression angle, 106, 109
Compression ratio, 106
Computational model, 25
CN feature extraction, 127
Critical band, 12
CV discrimination
    with synaptic adaptation, 73
    with two-tone suppression, 94
Dominant frequency principle, 6, 63
Dorsal cochlear nucleus (DCN), 121, 124
Dynamic adaptation, 57, 104
Dynamic (delta) parameter, 34
Dynamic rhyme test (DRT), 7

ERB scale, 14
Equal loudness coefficient, 102, 167
Equal loudness curves
    See Fletcher-Munson curves
Expectation maximization, 43

Finite impulse response (FIR) filter, 71
Fletcher-Munson curves, 19, 102

Forward masking, 21
Frequency discrimination coefficient, 94

Greenwood formula, 12

Half-wave rectification, 19, 60, 90
Hidden Markov model (HMM), 7, 38
Higher auditory system
    (See Central auditory system)
HMM recognition with
    asymmetric compression, 112
    CN features, 129
    psychoacoustic spectral subtraction, 153
    synaptic adaptation, 79
    two-tone suppression, 95
Hodgkin-Huxley model, 14

Infinite impulse response (IIR) filter, 66
Inner hair cells, 10

Language modelling, 26
Lateral inhibition, 23
Likelihood ratio, 42
Linear prediction, 2, 32
Linear discriminant analysis, 77
Loudness level, 17

Masking
    See Auditory masking
McNemars's test
    See Statistical significance test
Mel frequency cepstral coefficient (MFCC), 32
Middle ear, 9
Mean (average discharge) rate, 119, 126
    coefficient of variation, 133
    envelope detection, 125
    noise consideration, 130

time constant, 132
Mel frequency scale, 32

Mean opinion score (MOS), 7

Neural discharge rate, 10
Neurons, 120
    primary-like, 120
    chopper, 121
    pauser, 120
Non-linear transformations in auditory system, 16

Onset neurons, 121
Outer hair cells, 11

Perceptual linear prediction, 33
Perceptual noise estimation, 146
Peripheral auditory system, 9
Phase-locking property, 16
Post stimulus time histogram (PSTH), 20, 68
Posteroventral cochlear nucleus (PVCN), 121, 124
Power law of hearing, 18
Psychoacoustics, 16

Rapid adaptation
    See Synaptic adaptation
Rate-place theory, 15

Seneff's GSD model, 57, 117
Scatter matrix, 77
Short-term adaptation
    See Synaptic adaptation
Short-time Fourier transform (STFT), 1, 28
Simultaneous (frequency) masking, 22
Spectral subtraction, 137
    minimum mean-square error (MMSE) method, 137
    spectral smoothing with minimum statistics method, 138

psychoacoustic model, 150

speech enhancement, 138

Spreading function, 144

Static compression in MFCC, 108

Statistical significance test, 79, 162

Synapse model, 14

Synchrony, 17

Synaptic adaptation, 20, 55

Temporal companding

See Companding

Temporal masking, 21

Temporal-place theory, 15

Two-tone suppression, 22, 85

UCLA-SPAPL CV speech corpus, 74

Viterbi decoding, 44

Vowel clustering, 75

Zero-crossing

algorithm, 60

with peak amplitudes (ZCPA) model, 64

rate, 29

variance, 66

variance with two-tone suppression, 91

Zhang-Carney AN model, 53

www.ingramcontent.com/pod-product-compliance
Lightning Source LLC
LaVergne TN
LVHW042334060326
832902LV00006B/163